Design and Technology

Senior Author:
Kathy Browning, B.F.A. Honours, M.F.A., B.Ed.
Visual Arts Honours Specialist

George Heighington, Honours B.A., B.Ed.
Industrial Arts Specialist
East York Collegiate Institute
Toronto, Ontario

Virgil Parvu, B.A., M.Sc.
Centennial Secondary School
Welland, Ontario

Douglas Patillo, B.A.
Maplewood Elementary School
Essex County, Ontario

McGraw-Hill Ryerson Limited

Toronto Montreal New York Auckland Bogotá Caracas Lisbon London Madrid Mexico Milan
New Delhi Paris San Juan Singapore Sydney Tokyo

Design and Technology

ISBN 0-07-549650-X

1 2 3 4 5 6 7 8 9 10   G   2 1 0 9 8 7 6 5 4 3

Printed and bound in Canada

Care has been taken to trace ownership of copyright material contained in this text. The publishers will gladly take any information that will enable them to rectify any reference or credit in subsequent editions.

Canadian Cataloguing in Publication Data

Main entry under title:

Design and technology

Includes index.

ISBN 0-07-549650-X

1. Industrial arts. I. Browning, Kathy.

TT165.D38 1992   600   C92-094675-5

Developmental Editor: Fred Di Gasparro
Senior Supervising Editor: Marilyn Nice
Permissions Editor: Jacqueline Russell
Production Editor: Claudia Kutchukian
Design and Illustration: VISU*Tron*X — James Loates
Typesetting: VISU*Tron*X — Jennifer Loates, Susan Calverley

Cover photograph of the Canadarm manoeuvring the Hubble telescope was supplied by Spar Aerospace Limited. Reproduced courtesy of NASA. Canadarm is the name given to the Shuttle Remote Manipulator System (RMS), Canada's contribution to the U.S. Space Shuttle.

This book was manufactured in Canada using acid-free and recycled paper.

Contents

Chapter 8: MATERIALS

Chapter 9: FABRICATION PROCESSES

Preface

*T*hroughout history, our ancestors thought and planned how to use the materials around them to survive and to make their lives more comfortable. Within a relatively short span of time, humans have evolved from the firepit to the microwave, from travel on foot to travel by car, from hand labour to robotics, and from clay tablets to microchips. This evolution was possible because people began to understand the complexities of their world and to develop creative ways to use these complexities. Today's industrial designers continue to creatively use their technical knowledge of materials and processes to design new products.

In *Design and Technology*, you are invited to develop your own designs. You will learn to apply to different situations a design process that involves identifying a need, brainstorming, writing a design brief, researching, planning, fabricating, and evaluating. It was our desire to provide a model that you can use to develop lifelong decision-making and problem-solving skills that will help you in every area of your life.

Design and Technology is divided into two parts. Part One: Discovering Design begins with an overview of the importance of responsible and safe working habits. The chapters that follow explore the design process, design considerations, and examples of design in action. Part Two: Research and Development explores structures, energy, machines, materials, and fabrication processes.

Each chapter begins with What You Will Discover, which summarizes the learning objectives for the chapter. Key terms appear in bold type in running text and are also listed alphabetically in Terms to Remember at the end of each chapter and in the Glossary at the end of the book. See for Yourself activity boxes appear throughout the chapters; these activities provide you with opportunities to apply the concepts you have just learned. Each chapter concludes with Points for Review, a brief summary of key points, and Applying Your Knowledge, a section of review questions and activities.

Career Profiles are another feature of every chapter. These profiles are stories of Canadians who are in the design and technology field — why they chose to enter this field and the educational and personal paths they took to get there. We sincerely thank all those people who kindly provided their stories.

The projects at the end of each Part are very important features of the text. With these projects, you will apply the knowledge you have gained and practise using the design process. Some projects include a list of required materials and the required fabrication steps; other projects leave more decision making up to you.

With the hands-on approach, you will be able to plan creatively, question thoughtfully, work safely, learn through active participation, and expand your knowledge in a gratifying process.

Kathy Browning
George Heighington
Virgil Parvu
Douglas Patillo

Part ONE

Discovering Design

Chapter 1

SAFETY

What You Will Discover

After completing this chapter, you should be able to:

- Use your knowledge of safety every day.
- Identify the appropriate clothing to wear when working with tools, machines, and materials.
- Use the appropriate safety precautions when working with tools, machines, and materials.
- Understand the hazards and exercise caution when working with solvents, dust, and fumes.
- Understand the importance of paying close attention to your teacher, to ensure safety for all.

Safety is an important part of everyday life and needs to be emphasized in this course. The safe use of tools, machines, and materials is every student's responsibility. A co-operative effort is necessary to maintain a safe working environment, in order to reduce hazards that may cause accidents and injuries.

Getting Started

Before you go to school every day, you make decisions about your appearance. Personal hygiene and dress are part of these decisions. On the days that you will be working with equipment and materials, you should double-check to ensure that what you are wearing will not cause you injury. You may be dressed safely for everyday activities, but working with equipment and materials requires much more care.

Consider the following:

If you have long hair, it can drag your head toward a moving or rotating blade. If you wear long earrings when using an electric drill, your earring can become wrapped around the drill bit, and you could be injured.

Remember to tie back long hair.

It is your responsibility to wear appropriate clothing and be aware of your movements when you work around machines. Follow these general guidelines:

* *Tie back long hair.* If a shop cap is available, it is best to wear one. Bring something, such as an elastic band, to tie back your hair.

* *Remove rings, necklaces, watches, and other jewellery,* and store them in a safe place. When working around machines, these articles present a serious hazard.
* *Do not wear loose clothing* such as ties or baggy shirts and sweaters. They can get caught in the machines and can drag you in.
* *Wear short sleeves.* A long sleeve can get caught in machinery. If you are wearing long sleeves, roll them up over your elbows.
* *Wear pants that do not have cuffs* to avoid anything hot or sharp falling into a cuff.
* *Wear shoes that are comfortable,* preferably with nonslip rubber soles so that you do not slip and fall.
* *Do not wear open-toed shoes.* You could drop a hot or sharp object on your toes, or get a sliver of material embedded in your foot.

Be Aware of How You Act

Listen attentively to your teacher's lessons and demonstrations. They will help you to gain the information necessary for completing your projects and will save you valuable time. Listen attentively to the instructions on how to handle the tools, machines, and materials safely.

Don't push, run, shove, or bump into others, because someone can get seriously hurt by the equipment. Where equipment is in specific areas, move around carefully, as you would in a gymnasium. Keep inside the safety lines on the floor. The lines indicate definite spaces you should be aware of (for example, around an eyewash station). A circle indicates a general area, allowing passage in and out. The coloured half of the circle faces the area of caution.

See for Yourself

Examine the clothing you are wearing and check for rings, necklaces, and other jewellery. List any items of clothing, jewellery, or accessories that could be dangerous when you are using machinery. Combine your list with those of two other classmates. With your teacher, make a class list of these dangers.

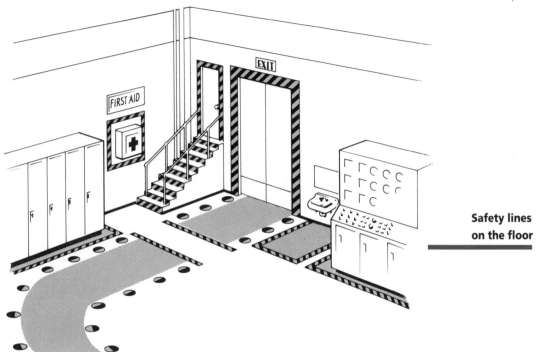

Safety lines on the floor

Disruptive behaviour such as making loud noises, screaming, or yelling will disturb the concentration of individuals who may be working on the machines. There should be only one operator per machine, and he or she should not be distracted. If you need someone's attention, simply walk over to his or her workbench and speak to the person, or wait until he or she has finished working on a machine. Concentration is very important — looking away from a machine for even a second while you are operating it could result in an accident.

Canadian Standards Association (CSA)

*T*he **Canadian Standards Association** (CSA) tests and certifies products used in the marketplace. The CSA symbol indicates that a product has met certain safety and performance criteria. The product has been tested and has met or exceeded the guidelines established for it.

All electrical products, such as electric drills, switches, and plugs, should bear the CSA symbol shown below. Safety products such as safety glasses, helmets, and hearing protection equipment must be approved by the CSA.

Look for this symbol of the Canadian Standards Association.

Wear Protective Gear

*M*ake sure you are protected for the task you wish to do. Just as people wear **protective gear** when they play sports, you too must wear protective gear when working with tools, machines, and materials. A downhill skier may wear goggles, or a hockey player may wear a face shield. You too may be required to wear goggles or a face shield when working.

There are several such protective devices, each to protect a different area of your body. Follow these general rules:

- *Wear eye protection at all times*. Safety glasses, goggles, or face shields must be worn at all times to protect your eyes when you are using any tool or machine or a material that can splash into your eyes. For example, a face shield is necessary when you are welding or soldering and when you are working with liquid finishes.

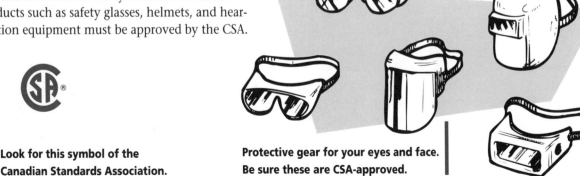

Protective gear for your eyes and face. Be sure these are CSA-approved.

- *Wear hearing protection at all times.* There are often noisy machines operating in a classroom. Sound-muffling gear, such as headphones, is necessary in a room where high-pitched machines are running. Exposure to noise levels above 110 dB over an extended period of time can cause permanent hearing damage.
- *Wear gloves when handling harmful materials.* This will prevent cuts when you are handling sheet metal and burns when you are welding, soldering, or handling hot plastic.
- *Wear a shop apron to protect your clothing.* An apron will help you to avoid staining, burning, or cutting your clothes. To avoid catching the apron strings in rotating machine parts, always tie them at the back — never in front of you.

Use Caution in Your Movements

Since you are constantly moving in a design and technology classroom, there is a greater potential for accidents. Therefore, you must be careful in your movements.

For example, when carrying heavy loads, use your common sense. When lifting heavy loads, bend your knees, crouch down, and keep your arms and load close to your body. Lift with your knees, and do not twist as you are lifting — keep your back as straight as possible.

Wrong way   Right way

Be sure to lift objects the right way to avoid back strain.

If the load is heavy, ask for help. When two or more people are lifting the same load, it is important that everyone work in unison. The load should be lifted at the same time by all carriers. The carriers should walk at the same cautious pace and put the load down gently at the same time. No matter how many carriers are involved, they should all move as if they were one person. If available, you can also use appropriate equipment, such as a chain hoist or roller, to help you.

Career Profile

Steve Copeland, Roger Ball, and Paradox Design

"Art is just a reward for doing well in the sciences." These are the words of industrial designer Roger Ball. They express the feeling he had while in high school in the sixties. Roger ignored that notion and kept drawing and building plastic models at home.

Steve Copeland got a different message from school and family. His father spent many hours with Steve building projects at home. Even his grandmother got involved, teaching him to design patterns and sew clothes. At school, his art teacher pushed him to draw.

Both these designers credit these early years with giving them the eye for three-dimensional design that they use today. Both went to the Ontario College of Art and received their degrees in industrial art.

Steve and Roger worked together at Cooper Canada, designing award-winning sports products with other design staff. In 1985, Steve fulfilled a long-time desire to open his own design company. Roger joined as a full partner the following year. In this way, Paradox Design was born.

Every design company has projects it is really proud of. Paradox Design focuses on sports products, because Steve and Roger really enjoy the design challenges these offer. The Itech® Concept II hockey face shield is one of many innovative products they have designed for Itech Sports Products Inc. in Montreal.

Steve and Roger also enjoyed the creativity involved in designing bicycles for Steve Bauer Bicycles of St. Catharines, Ontario.

In this project, colour, design concept, and even a theme for the bikes all interplayed to form the final products. Expanding their design products to include mobility aids in the late eighties was a great way to explore new ideas. Steve and Roger developed the Spirit® Power Chair with the manufacturer, Fortress Scientific. This gave them a greater scope of projects and design opportunities. Today, their design expertise is recognized across North America. Paradox Design is now seeking design projects around the world.

QUESTIONS AND EXERCISES

1. What were the three design considerations that Roger Ball and Steve Copeland used when creating their Bauer bicycles?
2. Why is it important to wear safety helmets like the one designed by Paradox?
3. What other sports can you think of that require wearing a safety helmet? List them.
4. Sketch five designs of ideas that you would like to create for the physically challenged.

Be very careful when carrying long materials. If you are walking down a hall with a long board, place a coloured cloth on the end of the board so that it can be easily seen. When carrying sheets of metal, wear gloves and cover the corners with cardboard or tape so that no one will get hurt.

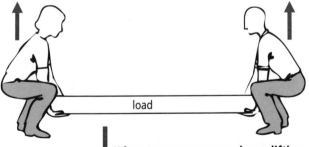

When two or more people are lifting, the load is lifted at the same time.

Carry sharp cutting instruments such as knives, saws, and chisels so that the cutting edge is away from your body and not pointed at anyone. Make sure the lids are on cans of stains, paints, and thinners. Keep the cans upright when you are carrying them.

Safe Use of Tools and Machinery

*Y*ou may be using power tools such as sabre saws, electric drills, belt sanders, pad sanders, vibrating sanders, grinders, and routers in the classroom. Safety guidelines, instructions, and parts lists are usually provided by the manufacturers. Be sure to read these instructions carefully before using the machines. As well, keep the following general guidelines in mind when using any tools and machinery, whether at home or at school:

• *Understand the safety instructions for each power tool before using it.* Your teacher will provide these.
• *Use a machine only after your teacher has taught you the proper safety procedures.* If you are uncertain about how to use a machine, do not hesitate to ask your teacher, who will be more than willing to show you. It is extremely important to understand the safety procedures in order to avoid accidents.
• *Choose the correct speed for the job.* Some machines operate at variable speeds or single speeds. Others are electronically controlled, in which case the speed of the machine is regulated automatically according to the material being cut. Since machines operate at different speeds, you need to choose the correct speed for the job. Your teacher will be able to advise you on this.
• *Always keep your hands a safe distance away from the cutting edge of a machine or tool* (for example, the blade or the bit), and never in front of it. Cut away from your body when using a knife, chisel, or saw. A sharp knife causes fewer accidents, because a dull knife can skip across the material being cut and injure the user.
• *Use tools for the purpose for which they are intended.* For example, do not use a chisel as a screwdriver.
• *Do not put small pieces of material on a machine.* They are too difficult to grasp safely. Small objects, such as pieces of dowel being drilled, may be better held in a holding device called a jig.

Using a jig with a drill press

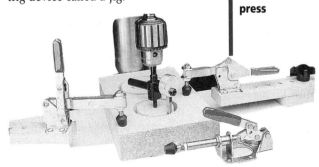

You may even be able to design and make a jig for others to use.

- *Shut off a machine when you are finished with it, and wait until it comes to a complete stop.* It is very important to make any adjustments on the machine only when the power is off. Be sure to pull from the plug at the power receptacle, not from the cord.
- *A properly adjusted machine is a safe machine.* Keep the machine tidy by cleaning it and picking up all your scraps before leaving it.
- *Wait patiently for your turn on a machine.* While waiting, you may be able to complete some other part of your project such as sanding or fitting, or checking your plan for the next step. Do not distract or interrupt anyone who is working on a machine. Stay outside the safety lines painted around the machine areas so that classmates who are operating machines can work safely.
- *Everything in the classroom has a place.* Remember the following rules:
  If you take it out, put it back.
  If you drop it, pick it up.
  If you open it, close it.
  If everyone follows these rules, it will be easier to find a tool or to clean up the shop quickly in time for the bell.
- *Report any broken or mislaid tools to your teacher.*

Safe Use of Solvents

*T*he major **solvents** that you might use are Varsol and thinners, plastic glue (ethylene dichloride, acetone), and methyl hydrate, or denatured alcohol. Flammable liquids must be stored in a metal cabinet. Spilled liquids must be wiped up immediately.

Varsol and thinners are used to clean the brushes that are used to paint or stain your work. Plastic glue is used to attach pieces of acrylic plastic to one another. Methyl hydrate is commonly used in shellac, which you may use to give a polish to your work.

Whenever solvents are used, your skin should always be protected, and you should always wash chemical splashes off as soon as possible. If your clothes become soaked with a solution, change them right away. Some solvents can be absorbed through the skin. They tend to remove the natural lubricants found in the skin, making it dry and possibly causing dermatitis, which is an inflammation of the skin. Skin-protection creams or disposable gloves can be used if this is a problem.

Eye protection such as safety glasses, goggles, or face shields should be worn if there is a possibility of solvents splashing into your eyes.

Workplace Hazardous Materials Information System (WHMIS)

*T*he **Workplace Hazardous Materials Information System (WHMIS)** is a Canada-wide system to ensure that employers and employees are provided with information about the hazardous materials they work with on the job, to protect their health and safety. WHMIS compels employers to (1) place warning labels on containers of hazardous materials and (2) make available separate safety-data sheets with detailed information. The information includes ingredients, fire and explosion data, health hazard data, procedures for handling spills and leaks, and special precautions related to hazardous materials.

WHMIS Classes and Hazard Symbols

CLASS A — Compressed Gas

CLASS B — Flammable and Combustible Material

CLASS C — Oxidizing Material

CLASS D

1. Materials Causing Immediate and Serious Toxic Effects

2. Materials Causing Other Toxic Effects

3. Biohazardous Infectious Materials

CLASS E — Corrosive Material

CLASS F — Dangerously Reactive Material

The Finishing Area

*I*n your classroom, a dust-free **finishing area** is usually provided. This is where you apply a finishing coat of stain, paint, shellac, or urethane to a project. Keep in mind the safety procedures when working in the finishing area.

The stains, paints, and chemicals used to finish a project can emit fumes that cause ear, eye, nose, and throat irritation. Make sure the exhaust fans are working or that windows are open. Handle the finishing materials with care in a well-ventilated area, and try not to inhale the fumes directly. Avoid direct contact of the materials with your skin. Read the instructions on each container.

METHANOL

– Liquid is flammable.
– Inhalation of vapour is harmful to health.
– May cause blindness if product is swallowed.

– Keep away from sources of ignition.
– In case of fire, use alcohol foam.
– Avoid prolonged or repeated breathing of vapour or contact with skin.

IF INHALED: Move victim to fresh air and perform artificial respiration if necessary. Contact physician.
IF SWALLOWED: If victim is conscious, give one glass of warm water containing one tablespoon salt in order to induce vomiting. Contact physician.
IF SPLASHED ON SKIN OR EYES: Flush affected areas with water for 15-20 mins.

SEE MATERIAL SAFETY DATA SHEET
ABC Company, 111 First St., Anywhere, ON

A typical supplier warning label

When projects are left to dry, be careful not to get dust or dirt on them or to brush against them as you walk by. When projects are completed, some of them may look the same. To avoid confusion, print your name or number in pencil on the back.

Rags or towels that have stain, Varsol, or solvents on them are a fire hazard and should be placed in an appropriate fireproof container. Brushes should be properly cleaned and placed in the containers provided. This keeps the area clean and tidy, ready for the next person.

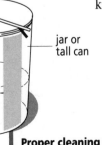

jar or tall can

suspend brush to protect bristles

Proper cleaning of brushes

First Aid

*I*f you or someone else is hurt, stop all work, remain calm, and report the accident to your teacher immediately. Even the smallest cut or scratch should be reported to your teacher and treated in order to minimize any infection that may arise. A careful record will be kept by the teacher in case additional information is required by the doctor.

You should be aware of the correct procedures to follow when an accident occurs. These will be provided by your teacher. You should also know where the first-aid kit is in case of an emergency.

Become familiar with the location and use of the eyewash station in the classroom. Any eye irritation should be attended to immediately.

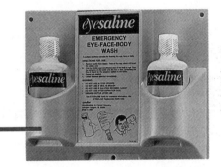

Eyewash station

In Case of Fire

*I*f a fire alarm is sounded, follow fire safety regulations as instructed. Shut off the power to the machines, lay your tools on a table, and proceed quickly and quietly to the designated exit. Staying calm will help everyone to get out quickly and safely. Be familiar with the fire escape routes in the school.

Fire Extinguishers

Under no circumstances should a student attempt to extinguish a fire! It is important to know the location of the fire extinguisher so that a trained person will know where to find it. The job of extinguishing fires is dangerous and complex, and must be left to trained people. Different types of fires need different types of firefighting equipment. For example, some fire extinguishers contain chemicals to stop fires fuelled by burning solids such as wood, but do not contain the chemicals for an electrical fire.

Check the fire extinguisher in your classroom for the information label. The label will have one or more symbols to identify which types of fires the extinguisher is designed for. What types of fires can be extinguished? Compare your findings with a classmate.

Ⓐ trash • wood • paper Ⓑ liquids • grease Ⓒ electrical equipment

Different extinguishers are used for different fires.

Industrial Accident Prevention Association (IAPA)

*T*he **Industrial Accident Prevention Association** (**IAPA**) is a government agency with offices across the country. The IAPA ensures safe practices in the workplace for practically all activities. It also distributes information on the treatment of accident victims and trains safety leaders for each workplace. As a student, you should make use of the IAPA's free information services for assignments.

Your Senses

*Y*our senses are your most valuable possessions. Take care of them, as they can be easily injured. Always stay alert — remember **A B C** (Always Be Careful) when working with equipment and materials.

Use your senses to your advantage. Watch for potential signs of danger, for example, a piece of wood or metal that might chip off and hit you in the eye. Listen to the sound of machines such as the band saw and the router. Does the band saw run smoothly as it cuts the wood? Does the router make a humming sound? Is there too much dust in the room, causing you to sneeze? Are there signs of an acetylene leak from the torch? Are there dangerous fumes? Do you smell something burning? Are there safety treads in front of a machine to prevent you from slipping? Is the floor in the finishing room slippery?

Do not touch machine parts that have been moving at high speed. A drill bit will still be hot after using it, even though it is no longer moving.

The Environment

*B*e aware of your environment, whether you are in a laboratory, a workplace, a school, or your home. Dispose of cans and bottles in proper receptacles for recycling. Waste paper and cardboard can also be recycled into other products. Sawdust can be used as a ground cover for flower beds or under bushes around the house. Scraps of wood or metal can be used to make abstract pictures or puzzles.

Always remember to recycle.

Keep the classroom clean and safe. The floor should be free from oil and grease. Sweep up metal chips and other waste materials.

Dust and Fumes

*W*henever possible, dust-collecting equipment should be used to remove the fine particles floating in the air when you are sanding, using a router, or cleaning a project, and also when you are cleaning the room. This will help students who are allergic to dust or who have respiratory problems. The ventilation devices provided on some machines should be used to take out dust from the air and away from cutting blades, bits, and sanding belts.

When welding or soldering, be aware of the materials that are being joined. Some fumes can cause severe reactions, such as flu-like symptoms. Check the ventilation system before starting work and, as you are working, make sure that the system continues to work well. Avoid inhaling the fumes, and wear a dust mask where needed.

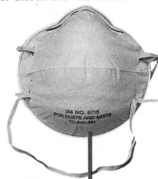

One of the dangers of burning or overheating plastic during welding or hotbending operations is the emission of vapours from the plastic. Do these operations in a well-ventilated area to avoid inhaling the vapours.

Dust masks will prevent you from breathing in harmful smoke and particles.

Signs and Symbols for Safety

*I*n your classroom, you will often find signs that encourage you to act safely. Often these signs will have no words, only symbols. For example, you may have seen the following sign on various containers:

This is the common poison symbol. It means that you should not ingest the material.

There are many other symbols that you will need to know. The signs that have common shapes are easy to recognize. A sign in the shape of a circle indicates a general requirement.

This symbol tells you that eye protection is required.

The square is used for information. These signs inform us of important facts, such as the location of an eyewash station.

Eyewash station symbol

A sign in the shape of a triangle indicates a warning. It alerts us to the presence of danger.

Electricity symbol

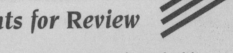

Points for Review

- Safety is an important part of everyday life.
- Wear appropriate clothing and protective gear, and consider your movements near machines and when using tools.
- Ensure your safety by paying attention to your teacher.
- WHMIS requires warning labels on containers of hazardous materials and information to be made available on these materials.
- Protect your skin, face, and eyes when using solvents.
- Use solvents in well-ventilated areas.
- Use the first-aid procedures outlined by your teacher.
- Know where to exit during a fire alarm, and follow fire-safety regulations.
- Keep your area clean and free of potential hazards.
- Use the appropriate ventilating equipment when necessary.
- Know the safety signs and symbols.

Terms to Remember

Canadian Standards Association
  (CSA)
finishing area

Industrial Accident Prevention
  Association (IAPA)
protective gear

solvents
Workplace Hazardous Materials
  Information System (WHMIS)

Applying Your Knowledge

1. Explain why listening attentively to your teacher is important in terms of safety.
2. Draw a set-up for keeping brushes clean and fresh.
3. Explain why it is necessary to use hand and power tools only after you have been instructed by your teacher.
4. Make a safety poster that uses symbols and words to do one of the following:
   a. provide information (for example, tell where the fire extinguisher is),
   b. present a regulation (for example, tell students to clean up their workplace), or
   c. serve as a warning of a potential hazard (for example, a student's inappropriate dress).

   When you are finished making a rough design, show your poster to a classmate. See if he or she can tell you what your poster is supposed to indicate. Ask your partner for advice on how your poster might be improved so that the message is more clear. Now make a final version of the poster.
5. Explain why projects must have a name or number written on them and be placed in a location where they are not in anyone's way.
6. What safety measures should you take when you are handling solvents?
7. Form groups of three to do the following exercise. The group members will be known as students A, B, and C. Using the information on safe dress in this chapter, student A will observe how student B is dressed and comment positively on any potential safety hazards; student B will observe and comment on student C; and student C will observe and comment on student A. (Remember that you can be critical without being negative.) Are there any potential dress hazards that were not mentioned in this chapter?
8. Explain briefly why eye protection must be worn when using tools and equipment and when in the finishing area.

Chapter 2

THE DESIGN PROCESS

What You Will Discover

After completing this chapter, you should be able to:

- Understand the role of designers and how they work.
- Identify the steps in the design process.
- Explore how the design process is used.
- Understand the importance of brainstorming in design.
- Explore different ways of researching as part of the design process.

*W*hen you want to design and build something, where do you start? What do you need to create, and why? What form will your project take, and what will be its function?

The process of answering these questions can be confusing, but anyone who builds something to solve a problem has in some way been answering these questions. Whether that person realizes it or not, she or he is using a method that has been used throughout the centuries. That method is the **design process**. Chances are you have already used this process without realizing it.

The design process outlined in this chapter will help to guide you through projects that you will undertake during this course. The process will enable you to tackle projects in an organized way that will lessen your confusion. You will find it rewarding to use the design process as an individual or a group activity.

The procedure is aimed at solving problems, which is something that you do every day. From the moment you wake up in the morning you are making decisions: What should I eat? What should I wear? What should I take to school? The design process will enable you to get started on a project and to complete it successfully. It will also help you with your future problem solving, for you will learn to think through situations, make assessments, and schedule tasks. The process also encourages you to take pride in ownership for your ideas.

These photos show one stage in the design process. The top one shows a computer drawing and the bottom one shows a clay model.

Designers

*T*he task of a **designer** is to develop a new or better way of creating something. Designers start from a **need**. The product to be designed may be something they need themselves, such as a musical instrument. The product may be needed by someone else, such as a special type of truck needed by a mining company.

All designers work through a similar process to find the best solution to a need. Designers have to be good problem solvers. They must be able to identify the need, consider all possible alternatives, develop a solution, and work on a plan through to completion.

Designers have to be aware of existing designs and changes in the design of a product. They can acquire this knowledge through research: they study the history of the product in books, magazines, newspapers, and computer databases. This helps designers to understand how many ways a product can be built. Designers may also attend conferences that present a variety of

Industrial designer Swanny Graham at work. Notice the different kinds of instruments she is using.

designs of a particular product. Good designers challenge themselves with each new design. They must be able to sketch a number of possible designs, and select the most successful version based on **form** (what the product looks like) and **function** (what the product is used for).

Designers invite comments, seek input from others, and make changes when and where necessary. For each product they create, they may have to go through the design process again and again before they are satisfied with the final product.

The Design Process

*T*he design process can be broken down into a series of stages. *There is no single design process that is right for every situation.* Following is a simplified version of the process, which can be changed in many ways.

Every time you start a project, you should go through these steps to clarify your ideas. If you get stuck at any one stage, you may have to "loop back" to the previous stage before going any further (see the broken line in the diagram). A detailed explanation of the design stages appears opposite.

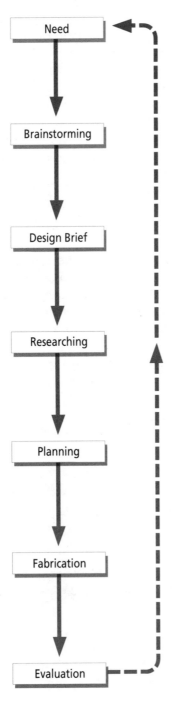

Stages of the Design Process

Need A need arises from a situation. From the situation comes the problem, but also an opportunity to fulfill the need by solving the problem.

Brainstorming Brainstorming is a quick means of generating ideas either verbally or on paper. It includes sketching, writing, speaking, and listening, in a group or individually. Brainstorming helps you to clarify the need and select the best solution.

Design Brief A design brief is a short statement explaining what you want to create, or fabricate.

Researching This stage includes gathering information, drawing several different designs, and making a model of the design selected. Discussions with your teacher will help you to select the best design.

Planning Planning involves deciding on materials, tools, and procedures to follow. You may need to prepare a bill of materials and use a chart to help you organize your time and the steps in your project.

Fabrication Fabrication is using the materials, tools, and procedures to construct an object that will fulfill the need.

Evaluation In this step, the fabricated object is tested in order to evaluate its success. Review the design process, and ask the following questions: Has the fabricated object fulfilled the need satisfactorily? Do I have to change or modify the design? How does the object compare with others that I have seen? What will I create next time?

Using the Design Process

*T*here is no mystery to using the design process — it is just an organized way to help you design. Using this process can help you find the best solution in the shortest time. The following sections explain each step of the design process in more detail.

Need

In order to design something, you must start with a *need*. This need may be one that you happen to recognize on your own. However, during your course you may be asked to *discover* a need. One way to do this is to ask yourself the following questions:

- What do I need to create?
- Do I need to build it for myself or others?
- What do I need personally?
- What do my family members or friends need?
- What do we need in the kitchen, family room, or yard?
- Where would the object be used?
- What could I build that would assist an activity and make it more easy or effective?

The more questions you ask yourself, the better. Often the answers to these questions will arise from situations. For example:

1. I have nowhere to keep my pens and pencils. They always seem to be lost when I need them.
2. I have difficulty doing my homework at my desk in the evening because the lighting is dim.
3. My family members tend to misplace their keys.

It is important to realize that these situations present not only problems, but also opportunities to solve the problems to fulfill the needs. The needs that follow from the situations described above are:

1. I need to make something to organize my pens and pencils.
2. I need to make something to improve the lighting in my room.
3. I need to make something to organize my family's keys.

Brainstorming

Brainstorming is the key to developing ideas. It encourages you to get your ideas out in the open, and it can be done as an individual or a group activity. Once you know the need, you can brainstorm solutions by writing or sketching all the possible alternatives.

Brainstorming can be a very effective individual activity, and it is also effective in groups as a quick way of generating ideas and sharing information. Remember that the most important part of brainstorming is the generation of as many ideas as possible. Later you can narrow down the possibilities. Following are some general guidelines for group brainstorming:

- Make sure everyone in the group knows what need is to be fulfilled.
- Allow group members to offer their ideas as quickly and with as little interruption as possible. Encourage creative, unusual, and even bizarre ideas. This is called "freewheeling."
- Withhold judgment. Accept all ideas, and evaluate them later. Sometimes the craziest idea will be the best.
- Write or sketch every idea that is suggested. At this point, quantity is more important than quality.
- "Spin off" from another person's idea. If someone's idea makes you think of another idea, **adapt** it. Combine or **modify** ideas to create new ones. It is all right to build on others' ideas.

In addition to spinoffs, you may want to use brainstormed ideas in the following ways:

abstract	— develop an idea in unexpected ways
eliminate	— delete an idea
magnify	— make an idea larger
minimize	— make an idea smaller
rearrange	— change the order of an idea
repeat	— make multiples of an idea
reverse	— change an idea to the opposite order
separate	— break apart an idea
substitute	— replace an idea with another

It is important to record your group's ideas in the form of notes, lists, descriptions, sketches, or computer drawings. **Sketching** is a rough form of drawing. It can include doodling and making rough notes (see Chapter 3 for more information on sketching). It is much easier to discuss a design with someone else when there is a visual representation to look at. That is why sketching is important at this point in the design process.

You should come up with a minimum of five solutions, because your first idea may not be the best.

Brainstorming sketches. At this stage, you are making decisions to narrow down the solutions.

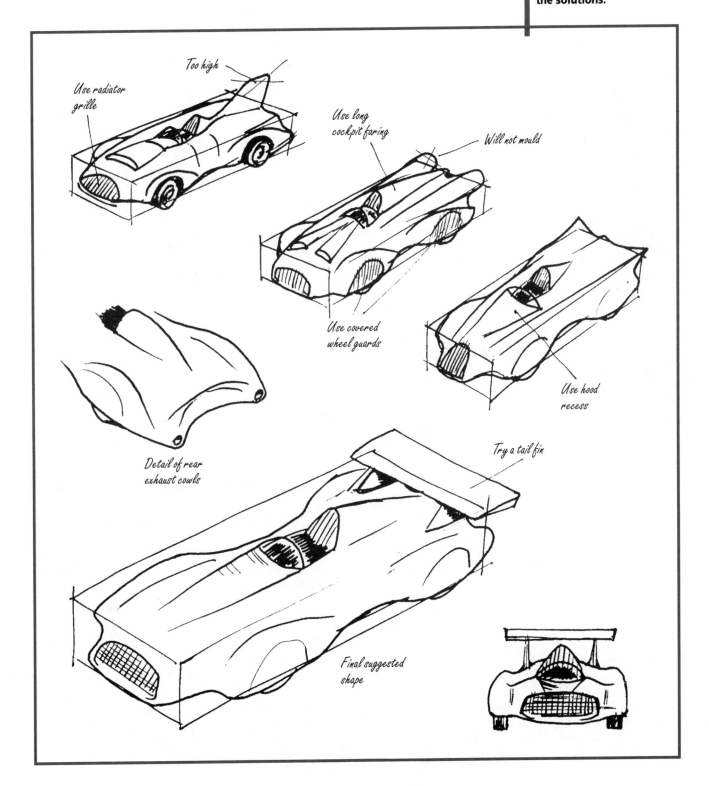

Use radiator grille

Too high

Use long cockpit faring

Will not mould

Use covered wheel guards

Use hood recess

Detail of rear exhaust cowls

Try a tail fin

Final suggested shape

One way to help you decide on a solution is to organize an **idea book**. Cut out designs and articles that you like from magazines and catalogues, and paste them into your idea book. Add your own notes and sketches from your brainstorming sessions. When you see all of your ideas together, they will help you to narrow down the possibilities when you are faced with a design situation.

An idea book

Once you have completed these steps, you are ready to have a conference with your teacher. Your teacher can help you to select the best solution. In your discussion with your teacher, you will want to consider the following important parameters: the materials, the tools, and the time available; the procedures to be followed; the size of the class, the room, and the object to be fabricated. These parameters will help you to decide which of the suggested solutions can be considered. Indicate on your sketches or list of ideas which ideas seem most appropriate based on the parameters that have been set. Consider a number of possibilities before deciding on a solution based on the information you have gathered. Your teacher will be able to advise you on whether you should pursue your selected idea. Once you have agreed on one particular idea, you are ready to write your design brief.

Design Brief

A **design brief** is a short statement describing what you need to create in order to solve the problem. You will have a greater chance of developing a successful design and project if you can briefly state the problem.

EXAMPLE 1
Need: I need a place to store my pens and pencils.
Brainstorming: (at least five possible solutions)
Design Brief: To design and create a unit to fit on my desk to hold pens and pencils.

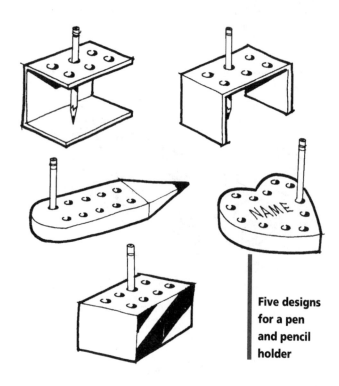

Five designs for a pen and pencil holder

EXAMPLE 2
Need: I need a better light source for my reading.
Brainstorming: (at least five possible solutions)
Design Brief: To design and create a reading lamp that will improve the lighting in my room.

EXAMPLE 3
Need: I need something to organize my family's keys.
Brainstorming: (at least five possible solutions)
Design Brief: To design and create a unit to organize our keys.

The design brief clarifies what you need to do in order to solve the problem. It helps you to remember your problem exactly. It also enables you to check, at a later stage, whether your final solution meets the need. Preparing a design brief is an essential part of the design process. Professional designers also use design briefs, which they pass on to **fabricators** if the designers are not making the objects themselves.

Once you have written your design brief, you are ready to begin researching: gathering the information you need to help you design the object that will solve the problem.

Researching

Begin your **researching** by asking yourself a series of questions. These should relate directly to your design brief and be specific to your project. Using Example 1, your questions might be:

- How many pens and pencils do I have?
- What are their sizes?
- Can they be organized by type, for example, H, HB, 2B?
- Where is the unit to be used?
- What design would work well with the objects that will be around it?

Write down your ideas as you ask yourself these questions. As you are thinking about these questions, you need to make at least five sketches that relate to your design brief. (You may or may not have made sketches during the brainstorming step, but here they are more important.) The sketches will be five possible solutions to your problem.

Narrowing down your designs

This one

You may produce new ideas or new versions of old ideas.

Many successful designs are forms of earlier designs. You may even want to combine several designs. Do not copy someone else's ideas — develop your own sense of design. Remember, too, that a design is not necessarily good just because it is different.

For each product that designers work on, they need to know about variations in existing designs and how they originated and developed. This is part of good research. It is important to look at the history of successful designs that relate to your project, and consider the different ways it can be fabricated.

Design Factors

Safety	• Will the object be safe to use? • How could it be made safer?
Function	• How will the object be used?
Aesthetics	• How will the object look? • For whom are you making the object?
Ergonomics	• How big and what shape must the object be? • What is the object's relationship to human size and form? • Will the object be comfortable, easy, and efficient to use?
Time	• How long will it take to make the object? • When is it needed?
Materials	• What materials do you need to make the object or complete the solution to your problem? • What materials are available? • What do you need to buy, and how much will it cost?
Fabrication	• What tools are needed to complete the job? Will you have to buy some of them? • Do you have the proper hardware to join the parts together? • Are you familiar with the fabrication processes? • What do you need to learn in order to complete the object? • What do you need to make the object work? • Do you want the object to move? • Do you want the object to create sound? • Do you need to know about simple circuits or magnets? • Do you need to know about other technologies, such as electronics or mechanisms?
Environment	• How will the fabrication process or object affect you or society?

Career Profile

Beverly Dywan and Museum Displays

"Teamwork is incredibly important in industrial design," says designer Beverly Dywan. "I work with objects, budgets, people, and environments, and must be able to handle all of these situations effectively.

"Kids, adults, physically challenged people, and seniors all must be taken into account when displaying things. The appropriate ergonomics must be taken into account. If you go into an exhibit and everything is too high for you to see, or the labels are written in a size that needs a magnifying glass to be read properly, it's uncomfortable. You will not enjoy yourself or learn anything."

Dywan remembers taking the first steps to becoming a successful designer as a child. She experienced a no-bounds approach to discovery and activity. "As a kid I really enjoyed making, drawing, and painting things," says Dywan. "I was making puppets and forts, and doing any type of craft that I could get my hands on, like papier-mâché or découpage or weaving. I loved painting my own furniture and enjoyed playing with my brothers' Meccano and Tinkertoy sets. In those days, girls weren't given such things, but luckily they were in my house and I'd get to use them!"

When Dywan attended secondary school, it did not offer art until grade twelve. "So from grades nine to eleven, I took art classes after school at community centres and with private instructors. When my school finally got art facilities, I spent extra time in the art room before and after school. I designed and produced posters for school dances."

After graduating from secondary school, Dywan studied fine arts at university. She took studio courses such as drawing, painting, and sculpture. Some of the most interesting work was making plaster casts of snow angels. She also studied courses such as psychology, sociology, and art history. "For me these courses were very important to my eventual career, because they helped me to reason," says Dywan.

Eventually, Dywan went to art college to study industrial design. She spent three years studying two- and three-dimensional design, drafting, computers, and drawing. She was able to create in wood, plastic, metal, and even food.

Dywan's specialty now is museum exhibit design as a principle designer at Design in Three Dimensions. The problems that clients propose are often similar. "They usually have objects that have stories to tell, and they want them displayed in a way that makes them interesting," says Dywan. Museum visitors must become interested enough to find out more about the objects and to read the label by a certain object. They may want to view a brief video on the display, or push a button to get information on a computer screen. Traffic flow has to be considered when designing exhibits so people don't get lost. Also, ergonomics has to be considered so that exhibits are displayed at a comfortable height, making labels easy to read.

"Museums are great places to learn things. The challenge of displaying objects also includes other criteria — like safety and conservation," says Dywan. "For example, most objects made of organic materials like paper, wood, or fabric must be kept under strict light, temperature, and humidity conditions. If you haven't noticed this, it's because the designers have done such a good job of making these technical criteria invisible.

"My favourite exhibit that I've designed so far was the one about baseball that was on at the Royal Ontario Museum for six months in 1989. It was a very large show, about 700 m^2, which is three times the area of an average house. It took one year to research, design, write, translate, and fabricate the baseball exhibit, and there was a core team of eight people working on it."

QUESTIONS AND EXERCISES

1. What situations does Beverly Dywan have to be able to deal with effectively?
2. List the ways that Beverly Dywan's education has assisted her career.
3. What ergonomic factors does Beverly Dywan have to consider when designing museum displays?
4. Design a museum display. For this you will have to state what the main idea is in your design brief. Sketch the outside dimensions of the display and the objects that will be displayed inside. Make a label that includes all of the information that you need to tell the public about your display.

It is also important to look at the appropriateness of the solutions that you have sketched. Consider the **design factors** outlined on page 17, as well as the design elements, such as shape, space, colour, and texture. (Some of these will be discussed in detail in Chapter 3. Others are discussed throughout this book.)

The best way to consider the design factors is to list them on a decision-making chart and rate each of the ideas that you have sketched. After you have calculated the totals, you will be better able to see which is the best idea, and why.

Suppose you have decided to create a better light source. The chart at the top shows how you might decide which design to use.

Decision-making chart for creating a better light source

RATE: 5 = excellent
4
3
2
1 = poor

IDEAS: (Alternatives)	Safety	Function	Aesthetics	Ergonomics	Time	Materials	Fabrication	Environment	Hold	Reject	Modify	TOTAL
Square Plastic Lamp	5	5	3	4	3	3	4	5			X	32
Hexagonal Plastic Lamp	5	3	3	3	1	2	3	5		X		25
Triangular Plastic Lamp	5	5	5	5	5	5	5	5	X			40
Round Plastic Lamp	5	2	2	3	1	2	1	5			X	21
V-shaped Plastic Lamp	2	1	1	1	3	2	3	5			X	18

The decision-making chart helps you to make an objective, unbiased evaluation in your decision-making process. However, subjective evaluation often also plays a part in the decision-making process. That is, you also listen to your intuition. **Intuition** is your "gut feeling" about a decision: what you feel is right.

Does your final decision based on an objective evaluation correspond to what your intuition tells you is the best design? If the answer is yes, proceed with your decision. Otherwise, you may want to rethink your decision. Talking to others to get their input may also help.

GATHERING INFORMATION
Make a list of the information you need and where to get it. This knowledge can be gained from first-hand experience through designing and building projects yourself. The more you build, the more you will learn. You can also gain knowledge from experts in the field.

In researching, you may need to contact professionals, such as industrial, graphic, theatre, and interior designers; artists; architects; scientists; mechanical, electrical, civil, aeronautical, and industrial engineers; tradespeople; technicians; carpenters; craftspeople; and suppliers. Besides your teachers, you are also likely to consult friends, family, and neighbours.

You can also acquire information from libraries, museums, and bookstores that carry magazines, newspapers, special trade publications, videotapes, audiotapes, and computer software. Local industries, colleges, and universities can also provide other material. You should keep notes on the information you find so that you can refer to it again later.

TESTING YOUR MATERIALS
You may have to decide which of the available materials is best for the fabrication of your project. The only way to decide this is to test your materials to ensure that the best one is used. For more information on materials and testing, see Chapter 8.

This student is researching in a library using a computerized location system.

ORGANIZING YOUR INFORMATION
While you are collecting information, you need to set up a system to organize it. There are several ways of doing this. Use a method that is suitable for your needs so that you can find your information quickly. Charts, graphs, maps, lists, and notes will help you to organize your data.

You need to collect, store, and sift through information while researching. Decide what is pertinent and what should be discarded, so that you are not bogged down with too much unnecessary information. The pertinent information — the information that relates directly to your project — will increase your knowledge of the subject. Teachers, family, friends, or professionals can help you edit your work.

It is better to collect too much material at this stage than not enough. You may need this information for a future project. After you have completed several projects, you can reorganize your material and discard some of it.

If you have a personal computer, you may want to set up files for your information so that you can store and retrieve it efficiently. Word processing programs will help you store and develop written information that you can use in a final presentation. Computer drawing programs will help you develop visual displays. *No matter what software you use, remember to keep backup disks!*

At this point, you need to **draft** your final design. Drafting is a precise and detailed form of drawing. You may also wish to make a model, which is a small version of your final project. Making a model will help you work out potential problems that you may encounter when building your project. Drafting and making models will be discussed in more detail in Chapter 3.

Planning

Once you have completed the steps in the research process, it is time to plan the fabrication of your design. **Planning** for a project will take into consideration the information you have gathered in your research: which materials will be best and how much you should use; the tools; the procedures; the time allowed; and the cost. (See

Steps to follow

Step no.	Steps to follow	Additional information
1	I need a better light source for my room.	Room is dark. Window is small.
2	Minimum of 5 sketches. Talk about my ideas with others.	Talk to family, friends, neighbours, teacher.
3	To design a lamp.	The lamp has to match my furniture.
4	Complete decision-making chart. Have conference with teacher. Make draft. Make model.	Which design is the best to make, and why?
5	Complete bill of materials. Complete chart of steps. Make template.	Ask teacher about materials. What colour of plastic do I want?
6a	Fabrication—cut wood for base of lamp using template.	How will I cut wood?
6b	File and sand edges of lamp base. Drill hole and organize lamp parts.	Talk to teacher about lamp parts.
6c	File and sand edges of plastic.	How do I cut the plastic?
6d	Bend plastic.	Talk to teacher about bending plastic.
6e	Use round file to put a groove in bottom of base.	I don't want base to rock on cord.
6f	Put finish on lamp base.	How long will it take to dry?
6g	Assemble lamp fixture.	How do I attach wires?
7a	Put in bulb. Put on lamp shade. Does it work?	Do I like my project?
7b	Written and verbal evaluation and presentation.	Which type of evaluation should I do?
7c	Review design process. Review chart of steps. What should I leave or change?	What did I learn?

Teacher's initials

Bill of Materials

Materials

Name of part	Material	No. of pieces	Thickness	Width	Length	Height
Lamp base	basswood	1	2 cm	30.5 cm	30.5 cm	
Lamp shade	plastic	1	0.25 cm		76 cm	25 cm

Tools

Name of tool	Class on use	Safety test passed
Crosscut saw	Sept. 17	Sept. 19
Files	Sept. 26	Sept. 27
Sandpaper and sander	Oct. 2	Oct. 5
Band saw	Oct. 14	Oct. 16
Drill and drill press	Nov. 1	Nov. 2

Hardware

Name of part	Material	No. of pieces	Size
Threaded rod	metal	1	Same size as drill bit, 2.5 cm lengths
Cord	plastic and wire	1	2 m length
Lamp switch	metal	1	Height with bulb doesn't exceed top of plastic.
Cord switch	plastic	1	Do I get this from the teacher?

Teacher's initials

the list of design factors on page 17.) Having gathered information on each of these areas, you can now plan knowledgeably.

Ask yourself these questions: What materials are best for making this object? What did my material tests tell me? How will the object be made? What tools do I have? Whether you are new to design or you are working in an established firm, you always have to work within the restrictions of time, cost, and availability of tools and materials.

At this stage, you may need to prepare a **bill of materials**. This is a list of the number of pieces of wood, plastic, metal, or other materials you need, including their sizes. As well, you will have to list the tools and hardware (e.g., nails, screws, glue) that you will require for

the project. You will also have to figure out what you need to buy.

Once you have all the necessary materials, tools, and hardware, list the fabrication procedures and think of how different parts of the object will be joined.

Making a chart that lists the steps to follow is a useful exercise. (See the example on page 20.) Check off the steps as you move through them. This will help you to keep track of what you are doing so that you can complete the project in the required amount of time.

If the process needs readjusting as you are completing the project, it is all right to try an alternative solution. Even though most of the adjustments should have been made at the model-building stage, you cannot foresee everything!

Fabrication

Sometimes a good design can be ruined if the item is not built carefully enough to withstand regular wear and tear. Engineers and architects who design bridges and highrise buildings have to be very careful about structures and materials in order to combat the stresses of weight, tension, and compression, as well as weather and natural forces. Consider these questions:

- Why does the chair you sit on need to withstand the stress of your weight?
- What would happen if a highrise building or a bridge was not able to resist the stress of an earthquake?
- How many types of materials could be used to build the structure of a highrise building or a chair?
- How could a highrise building or a chair be improved structurally?
- What are some considerations for making a comfortable chair?
- Other than skis, skateboards, and surfboards, what other objects need to deal with downward load?
- Why is corrugated cardboard stronger than a flat piece of cardboard?

You need to consider suitable materials, methods of fabrication, and surface finishes. You need to know the functions of each tool and how to handle each safely. Ask yourself these questions with regard to **fabrication**:

- What materials are most suitable and what sizes are required?
- What type of structure should be used?
- How should materials be joined?
- What tools and machines are required?
- What fabrication processes should be followed?

The first completed object is used to test and verify the design. Testing for strength and durability is very important. If the object is not safe, the design is not successful.

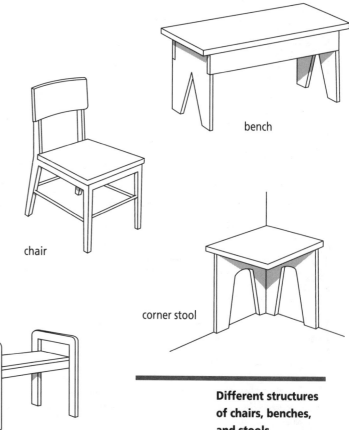

bench

chair

corner stool

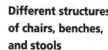

stool

chair

bench

Different structures of chairs, benches, and stools

See for Yourself

One of the best ways of testing bridges is to build a variety of bridge designs using straws or spaghetti, and glue. Build the bridge design of your choice and test its strength. Apply force by putting the bridge in a vise. As you test the bridges made by you and your classmates, you will be able to see which method of building the bridge was most successful by observing which one holds up the longest. Record in your notebook the material that seemed the strongest, and why. What is the relationship between structure and materials?

For professional designers, the first fabricated design is a prototype. A **prototype** is an actual-size product. It goes through an elaborate process of testing and evaluation before being used as a prototype for **production**. This is the time to test the joinery, or places where materials are joined. It is important that the joinery be strong — if it is not safe, the product will not be successful. You should have the design approved by your teacher for safety, ergonomic, and environmental factors before you proceed to create multiples.

Remember that it is difficult to achieve perfection the first time you make something, but it is important to do the best you can. (See Chapter 8 for more details on materials and Chapter 9 for more details on fabrication.)

Evaluation

In your **evaluation**, it is important to go back and review the stages of the design process. Your evaluation should involve answering the following questions:
- Did the object satisfy the need?
- Does your solution match your design brief?
- Did it solve the problem?

- Was your research accurate and appropriate?
- Did you create the most suitable design?
- How is your design different from or similar to others that are used for the same purpose?
- Can your design be adapted for other uses?
- How does your object affect the environment?

You may need to make a few improvements. List or sketch other possible solutions to the problem. What should stay the same and what needs to be changed? Should you have made changes at the planning and fabrication stages? Were the materials used the best for the object? Did you stay within your time frame and budget?

Your work should be evaluated by yourself and others verbally or in writing, including charts and comments. Evaluation is important because it refines your work and helps you to develop a sense of design and a critical eye for assessing projects. It helps you to improve your design the next time. Always analyze a project objectively. This means looking at the project for what it is and what it can or cannot do. It is best not to be too personal about what you have created — try to put yourself in the shoes of someone who is trying your project for the first time.

At this stage, you will make a final presentation of your project. You will present your sketches, drafts, model, lists, notes, charts, and the final project for assessment by your teacher and classmates. A presentation can take many class sessions, as this is an important part of the design process.

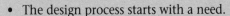

Points for Review

- The design process starts with a need.
- The task of an industrial designer is to create a new or better way to satisfy a need.
- The design process is a problem-solving method.
- There are several steps in the design process: determining a need, brainstorming, preparing a design brief, researching, planning, fabricating, and evaluating.
- Consider the design factors (safety, function, aesthetics, ergonomics, time, materials, fabrication, and environment) and design elements (such as shape, space, colour, and texture) in your solution to a problem.

- Begin your researching by asking yourself a series of questions relating to your design brief and sketching at least five solutions.
- Gather as much relevant information as possible, organize your information, plan your materials and tools, prepare a bill of materials, and make a chart of your fabrication process.
- In your fabrication, use suitable materials, tools, methods of fabrication, and surface finishes.
- Evaluate projects verbally and in writing by reviewing the stages of the design process.

Terms to Remember

abstract	draft	intuition	rearrange
adapt	eliminate	magnify	repeat
bill of materials	evaluation	minimize	researching
brainstorming	fabrication	modify	reverse
design brief	fabricators	need	separate
designer	form	planning	sketching
design factors	function	production	substitute
design process	idea book	prototype	

Applying Your Knowledge

1. Briefly explain the purpose of the design process.
2. Why do you think the design process starts from a need? Write down a few reasons, and then discuss them with a partner.
3. Why is it important to have a design brief? List ways that the design brief helps you in your designing.
4. In a brief sentence, explain each of the group brainstorming guidelines on page 14.
5. During the school year, you will probably be required to go through the design process with the goal of completing a fabricated object.
   a. List one possible need in your home.
   b. Brainstorm solutions to this need by yourself.
   c. Choose one solution and write a design brief for it.
   d. List the possible resources you could use to obtain information for your planned solution.
6. Complete a decision-making chart for the solutions (objects) you brainstormed in question 5b.
   a. Which idea received the highest score?
   b. Explain why this is or is not the object that you would choose to create.
7. Complete a bill of materials for the object you selected in question 5c.
8. Explain why making a chart of the steps to follow is important when creating your project. (See page 20.)
9. Complete a chart of the steps to follow for the object you selected in question 5c. Discuss with a partner the differences and similarities between your charts.
10. How will the availability of time and materials affect your project? Explain in a short paragraph.

Chapter 3

DESIGN CONSIDERATIONS

What You Will Discover

After completing this chapter, you should be able to:

- Identify some of the considerations that affect design.
- Understand how design is communicated through drawing.
- Explore the techniques used in sketching and drafting.
- Understand the importance of models.

*I*n Chapter 2 you discovered a design process that can help you to collect and organize information to solve a problem. By using this process, you can design a product to satisfy a need. In order to use the design process fully, you need to understand design considerations. You also need to develop good drawing techniques to represent these considerations.

Important Design Considerations

*I*ndustrial designers must be able to create products that will satisfy consumers. Designers do this by considering **design elements**: shape, space, colour, and texture. The design elements have a direct influence on the appearance and appeal of products.

Before deciding on the best design, designers also need to consider design factors: safety, aesthetics, function, ergonomics, time, materials, fabrication, and environment (see page 17 for a list of design factors). In the following sections we will discuss ergonomics; the other design factors are discussed in detail in other parts of this book.

Shape

A **shape** is an area that is defined by a line. It can be a two-dimensional plane or a three-dimensional form.

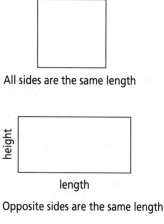

All sides are the same length

Opposite sides are the same length

Different lengths on all sides

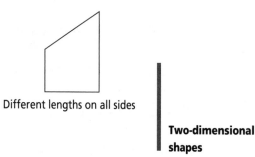

Two-dimensional shapes

A **plane** is a flat surface. Shapes have two or three planes. A two-dimensional shape has two planes: length and height.

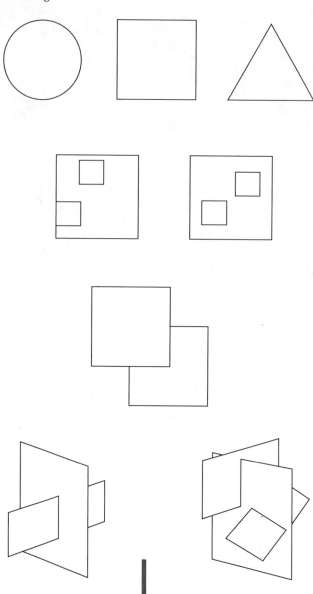

Planes

A **form** is a shape that has three dimensions: length, width, and height. An example is a cube. The form of a design is influenced by the function and purpose of the design.

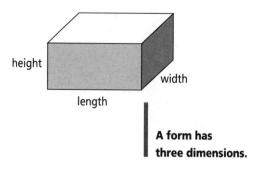

A form has three dimensions.

Space

Space is the area that surrounds an object or is contained within an object. The space is defined by the lines that encompass the object.

One way to clarify space is by understanding **balance**. There are two types of balance: symmetrical and asymmetrical. **Symmetrical balance** means that the two sides of a shape or form are the same or are mirror images of each other. **Asymmetrical balance** means that the two sides are not the same.

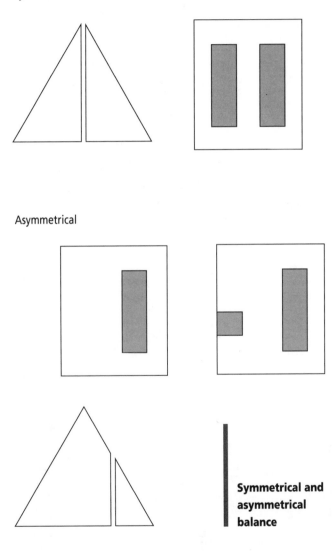

Symmetrical

Asymmetrical

Symmetrical and asymmetrical balance

Colour

Colour is important in design because it produces an instant response. The colour of an object should work with the colours of where it is to be used. For example, a red plastic lamp may not fit in with green leather furniture.

Different colours within an object also have to work together. Good designers make use of colour to enhance design. Often a natural colour, such as the colour of wood, will achieve the desired effect.

There are three **primary colours**: magenta, yellow, and cyan. These colours contain no traces of other colours. **Secondary colours** are created by mixing two primary colours.

It is important that designers know how colours work. For example, dark colours make objects appear smaller than light colours do; magenta, red, and yellow are warm colours; green, cyan, and blue are cool colours. To think of warm colours, think of the colours of the Sun. To think of cool colours, think of the colours of water.

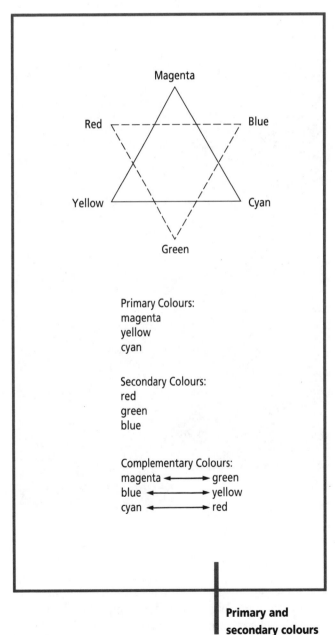

Primary Colours:
magenta
yellow
cyan

Secondary Colours:
red
green
blue

Complementary Colours:
magenta ⟷ green
blue ⟷ yellow
cyan ⟷ red

Primary and secondary colours

Texture

Texture affects the feel and appearance of a design. Texture is to the sense of touch what colour is to the sense of sight. Nature may give you ideas about texture. For example, pebbles on the beach might make you think of making a stone tabletop. A leaf might make you think of burning a design into your fabricated project.

Different textures

Ergonomics

Ergonomics is the relationship of an object to human size and form. In designing consumer products, designers place a lot of emphasis on ergonomics.

Ergonomics has an important bearing on the function of a design. Designers must be concerned about human safety and performance. They take into account human size and form to ensure that a design is suitable and

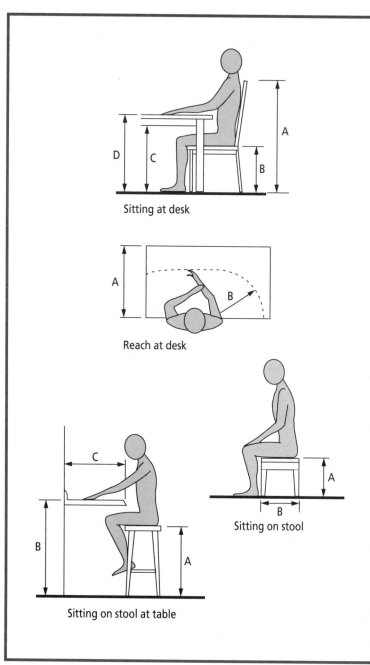

Sitting at desk

Reach at desk

Sitting on stool at table

Sitting on stool

effective for human use. In considering ergonomics, designers need to address these questions:
- How and where is the product to be used?
- How big and what shape will the product be?
- Will the product be comfortable and easy to use?
- Will the product be efficient?
- Can the product be used safely?

The dimensions of each part of a design must be considered in terms of human function, comfort, and efficiency. You may want to consider measuring the dimensions of these parts if they relate to your design.

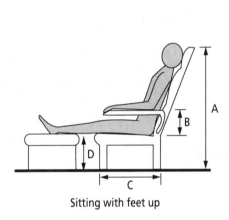

Sitting with feet up

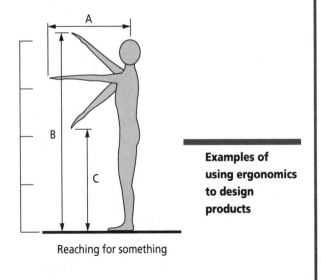

Reaching for something

Examples of using ergonomics to design products

Career Profile

Jonathan Crinion and Crinion Associates

"The designer's role is constantly changing," says industrial designer Jonathan Crinion. "Because designers are in a profession in which they are creating things, it is important that they are also socially responsible. It is important to be able to see whatever the designer creates in the context of the work and the contribution it will make."

Crinion's interest in design began when he was young. His father was an architect and his mother was a clothes designer/maker. "My father was always working on an architectural project, which our family would all discuss the merits of," says Crinion. "And my mother would be fixing something or coming up with a new design for a piece of clothing. Somewhere I got a curiosity about how things worked. I used to take everything apart to see how it worked — I just had to know."

His schooling was varied; he accumulated courses at college and university, and then the Ontario College of Art, where he graduated with an honours degree in product and systems design.

He decided to postpone his education for a time so he could work and travel. During that time, Crinion worked for Parks Canada in western Canada, where he learned to design pictograms that were silk-screened. "I developed some pictograms that you now see in parks across Canada."

Upon graduating, he worked with a design consulting firm. He worked on designs for the Royal Ontario Museum and designed all sorts of products, such as chairs and ovens.

Gazelle chair

In 1985 he founded Crinion Associates, in Toronto, which designs mainly electronic products and systems furniture. Crinion went on sabbatical for a time to work in London, England for a well-known architect, Sir Norman Foster. There he worked with 200 other designers.

Now back with his company, Crinion stresses that design should not only fulfill needs but also show concern for our environment. "Concern for the environment and the use of resources are new challenges for the designer," says Crinion. "Ask yourself if the solution is an object or a different way of doing things. Much of design is the design process itself. The solution is not always a product."

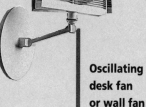

Oscillating desk fan or wall fan

QUESTIONS AND EXERCISES

1. How did Jonathan Crinion's parents affect what he is doing today?
2. List the ways that adults might influence your future career.
3. What are some of the design considerations that are challenges for Jonathan Crinion?

Drawing and Design Considerations

*Y*our design considerations should be included in a drawing. Such a drawing will remind you at a glance what you are making and will enable you to fabricate your project accurately.

Sketching

You have already discovered in Chapter 2 that sketching allows you to clarify your ideas in both brainstorming and researching. Sketching is used to convey ideas quickly. You will also need to know how to sketch different views of a design.

The pencil is one of your most important drawing tools. You will need to use a variety of pencils to draw different lines and to achieve tone. Pencil leads come in several grades: H is a hard grade and B is a soft grade; there are also different grades of hardness and softness. H pencils will produce thin lines. B pencils will produce thicker lines. 2H pencils are very hard and are used for faint lines and detailed work. 2B pencils are very soft and are used for shading and tone. HB pencils are "medium" — they are used for general work. No matter what type of pencil you use, always keep it sharp.

Different pencil grades and their uses. Shapes are drawn with 2H pencils. Forms are drawn and toned with 2B pencils.

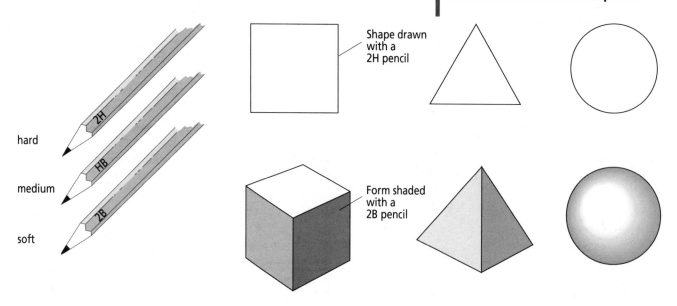

hard

medium

soft

Shape drawn with a 2H pencil

Form shaded with a 2B pencil

You will be able to sketch lines and shapes well with a little practice. Hold your pencil lightly and keep your wrist relaxed. Practise drawing lines by joining dots. Work on **horizontal**, **vertical**, and **diagonal** lines as shown in the diagrams that follow. Try to make the lines thicker as you go along. Then use these lines to create shapes.

When you sketch large circles, keep your wrist firm and make a circular movement with your whole arm. On a piece of paper, practise drawing circles and **ellipses** as shown in the diagrams below.

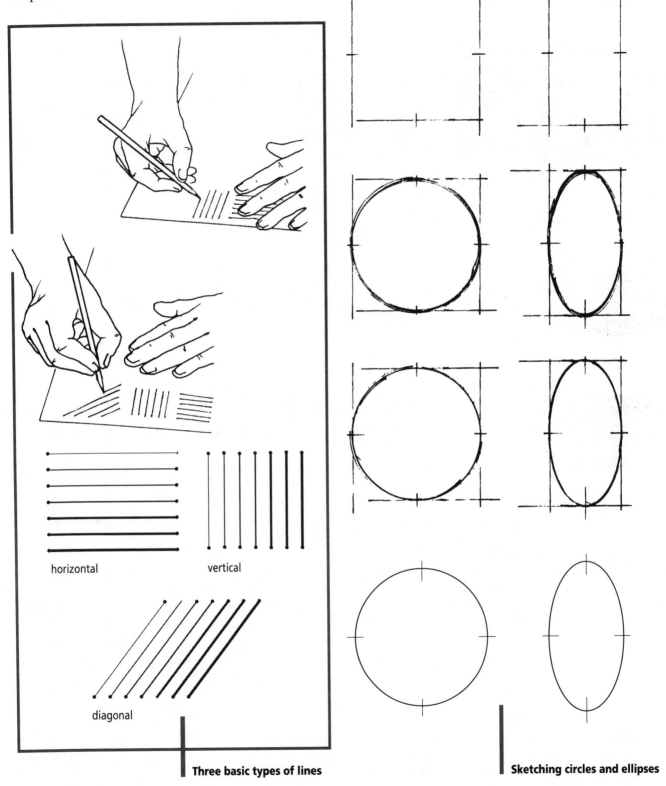

horizontal

vertical

diagonal

Three basic types of lines

Sketching circles and ellipses

Using basic lines to sketch an airplane

1. Sketch basic shapes.   2. Sketch tail fins.

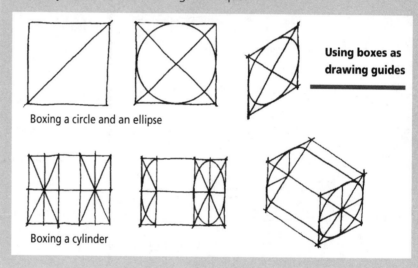

Knowing how to box a drawing will help you to develop your drawing skills. Use a box as a guide to sketch a circle, an ellipse, or a cylinder. With a sharp pencil (H or HB), lightly sketch a box. The box acts as a guide to assist with the overall proportions of the object being drawn. Now sketch the circle, ellipse, or cylinder to fill the box (see the diagrams below). Erase the box lines when the drawing is completed. As you become more skilled, you will be able to draw more complicated designs. Be sure to keep these drawings in your notebook to remind yourself of this drawing technique.

Using boxes as drawing guides

Boxing a circle and an ellipse

Boxing a cylinder

Drafting

Besides sketching, you will need to do some **drafting**, which is a more technical and accurate type of drawing. In fact, in many cases you will first have to make sketches, and then use the best sketch as a guide to make a detailed draft. Drafting involves the skillful use of pencils and drawing instruments, and it is the most precise form of detailing information about fabrication. It is important to make a detailed draft of the selected design so that you have an accurate plan to work from.

ISOMETRIC PROJECTION
An **isometric projection** is a drawing that shows the length, width, and height of a three-dimensional object. An easy way of learning how to draw an isometric projec-

tion is by drawing a set of steps. The following instructions will help you.

Step 1: Using a set square, draw the front view of three steps. The "front view" is often not the front of the object, but the view that gives you the most information about the shape.

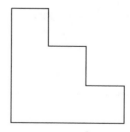

Step 2: With a protractor, measure an angle of approximately 30° from the bottom line.

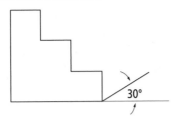

Step 3: Move your set square or ruler up the stairs and draw a series of parallel lines.

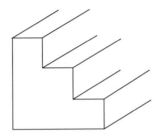

Step 4: "Walk" your way up the stairs by drawing vertical and horizontal parallel lines.

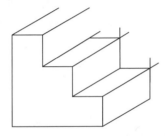

Step 5: Complete the set of stairs.

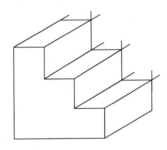

Step 6: Erase the extra lines. Add the length, width, and height dimensions to your isometric drawing.

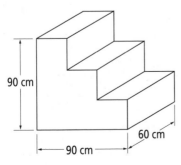

PERSPECTIVE

If you look down a street, the far end looks much smaller than the end that is nearest to you. When you look at a long block of wood, the far end looks smaller than the end that is nearest to you. This is because of **perspective**: the effect of distance on the appearance of objects. An important part of perspective is the horizon line. The **horizon line** is your eye level on the horizon, and it shows you the "vanishing point" of the street.

In one-point perspective, there is one vanishing point on the horizon line. In two-point perspective, there are two vanishing points. One-point perspective is the simplest way of showing perspective, whereas two-point perspective is more advanced and is helpful in developing drawings.

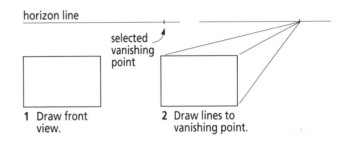

1 Draw front view.

2 Draw lines to vanishing point.

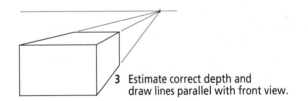

3 Estimate correct depth and draw lines parallel with front view.

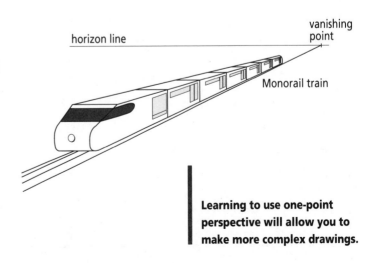

Monorail train

Learning to use one-point perspective will allow you to make more complex drawings.

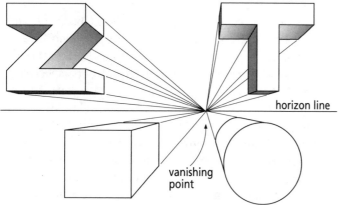

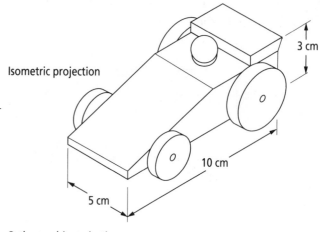

Isometric projection

Perspective is an important concept in industrial design and architecture. It helps you to understand an object in relation to where you are, and makes your drawings look more realistic.

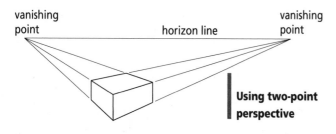

Using two-point perspective

ORTHOGRAPHIC PROJECTION

An **orthographic projection** shows the front, top, and side views of an object separately. The labels "Front" and "Side" go below the drawings, and "Top" goes on top.

The front view, which always gives you the most information about the shape, is drawn in the bottom left-hand corner. Use the front view to help you draw the side and top views. Then label the three views. After labelling, you should be able to see the three planes in the side and top views.

As you progress through your design and technology course, you will be able to draw more complicated isometric and orthographic projections like those shown in these diagrams.

Orthographic projection

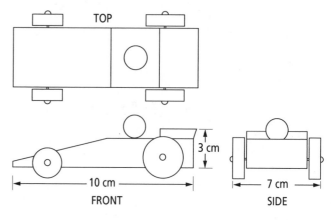

DIMENSION

Dimension is a measurement of length, width, or height. Our world is three-dimensional. The room that you are in, the school, the city, and the province are all three-dimensional. Three-dimensional objects have length, width, and height.

Objects take up space. Each object has a particular shape and form. One way of looking at objects is to know their outside dimensions (OD) and their inside dimensions (ID). Some forms need to be considered for both their outside and inside dimensions.

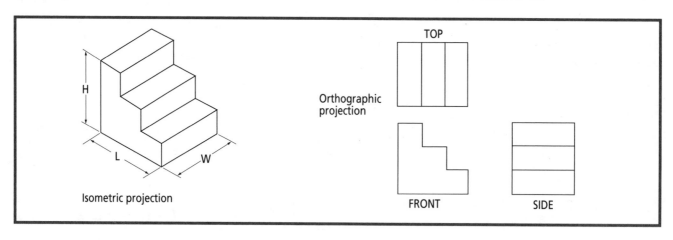

Look around your classroom for objects that you think should be considered for their inside dimension. To determine this, ask yourself, "Does this object contain something? Does the inside have to be a certain size?" Make a list of these objects. Discuss your observations in a small group.

A box that will contain something will have an ID that is important. The OD will be equal to the ID plus the thickness of the material used.

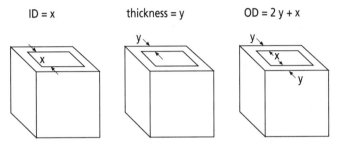

ID = x   thickness = y   OD = 2 y + x

How outside dimensions and inside dimensions are related

It is important to establish the dimensions of your project and provide exact metric measurements on your draft. Fabricated objects have an OD and ID. However, in drafting these are stated as overall dimensions and detail dimensions. **Overall dimensions** indicate the overall length, width, and height of a three-dimensional object. **Detail dimensions** indicate the sizes of any contours or details of an object, other than the overall length, width, and height.

Extension lines are thin black lines that extend from the lines that define the object. **Dimension lines** are thin black lines that are used to indicate a measurement.

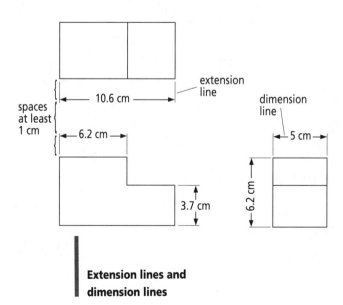

Extension lines and dimension lines

They span the distance between pairs of extension lines and have an arrowhead at each end. The arrowheads must touch the extension lines. When possible and practical to do so, place the dimensions above the front view. Always show the full-size dimensions of an object, regardless of the scale of the drawing.

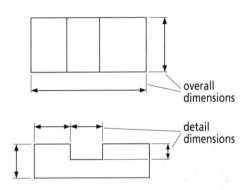

Overall dimensions and detail dimensions

COMPUTER-AIDED DESIGN (CAD)
Designers use **computer-aided design (CAD)** to make extremely detailed drawings of their projects. If you have access to a computer with a CAD program, you should use it to do your drafting.

The CAD program assists you in seeing and drawing your ideas. It allows you to rotate the draft of the object so that you can look at it from all sides. You will be able to see your draft from any angle, reduced or enlarged, modified or rearranged. The computer can also print many copies of your multiple-view designs. You can then use the copies for adding notes, colour, and other information where needed.

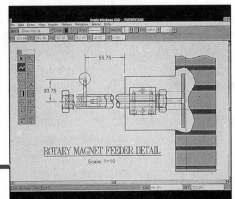

Using CAD to draw a design

Models

Using your drawings, you can make a model of your project. A **model** is a small version of your project that helps you to work out any problems that may arise while fabricating your final design.

A model can be made of cardboard, paper, straws, popsicle sticks, Plasticine, balsa or other materials. Cardboard and bristol board are excellent materials to use because they can be cut, bent, glued, fastened, punched, painted, and formed.

Model airplane

If you make changes during your planning, you may have to build several models of your design. Industrial designers and architects often make models of their ideas; models are especially effective for making presentations.

TEMPLATES

You may want to make a type of model called a template. A **template** is a pattern that you use as a guide to trace the shape of your project onto your material. The pattern is often cut out of paper or cardboard. Templates will allow you to cut several pieces of material that are the same shape and size. They are very useful if you need to make several models of a design.

Points for Review

- Designers must consider the design elements: shape, space, colour, and texture.
- Shape is an area that is defined by a line. It can be a two-dimensional plane or a three-dimensional form.
- Space is the area that surrounds or is contained within an object. Space may have a symmetrical or an asymmetrical balance.
- Colour enhances design. The primary colours are magenta, yellow, and cyan. Secondary colours are created by mixing two primary colours.
- Texture affects the feel and appearance of a design.
- Ergonomics is the relationship of an object to human size and form.
- Drafting involves the skillful use of pencils and drawing instruments to show detailed information about the design to be fabricated.

- An isometric projection shows the length, width, and height of an object.
- Perspective helps you to understand an object in relation to where you are and makes your drawings look more realistic.
- An orthographic projection shows the front, top, and side views of an object separately.
- Overall dimensions indicate the overall length, width, and height of an object. Detail dimensions indicate the sizes of any contours or details of an object other than the overall dimensions.
- CAD stands for computer-aided design. With CAD, you can draw and see objects on a computer, and rotate, reduce, enlarge, modify, or rearrange them.
- A model is a small version of a product that helps to provide insight into problems that may arise during fabrication of your design.

Terms to Remember

asymmetrical balance	dimension	horizontal	secondary colours
balance	dimension lines	isometric projection	shape
colour	drafting	model	space
computer-aided design (CAD)	ellipses	orthographic projection	symmetrical balance
	ergonomics	overall dimensions	template
design elements	extension lines	perspective	texture
detail dimensions	form	plane	vertical
diagonal	horizon line	primary colours	

Applying Your Knowledge

1. List the three dimensions in a form.
2. a. What are the two types of balance?
   b. Draw an example of each type of balance.
3. a. List the warm colours and the cool colours.
   b. Do you agree with these classifications of colour? Write a sentence at the bottom of your list telling why you agree or disagree with these classifications.
   c. Give an example from nature of each colour.
4. Make a list of all the ways you think colour can affect design. Discuss your ideas in a group.
5. Divide a piece of plain paper into six sections. Place this paper against different textured surfaces, and make rubbings of six different textures.
6. a. Explain briefly what ergonomics means.
   b. How does ergonomics affect the design of an object? (HINT: Think of the chair you are now sitting in. Is it comfortable? Does it fulfill its intended purpose?)
7. List the different grades of pencils and describe what they do.
8. Choose an object in your home or in the classroom that you would like to redesign. Sketch five new designs of that object.
9. Think of something you may need in your room at home. What would fulfill that need? Sketch five possible designs of this object.
10. Draw an isometric projection of your object from question 9.
11. Draw an orthographic projection of your object from question 9.
12. How does perspective assist you in drafting?
13. a. Explain overall dimensions and detail dimensions in a brief paragraph.
    b. Why do you need to know OD and ID when creating a container?
14. a. What is CAD?
    b. List three ways in which it assists design. Discuss your answers with a partner.
15. In a group, discuss why you should construct a model of your project.

Chapter 4

DESIGN IN ACTION

What You Will Discover

After completing this chapter, you should be able to:

- Understand the uses of design in two important fields: aerospace technology and musical instruments.
- Investigate how sound is created.
- Show that the laws of nature can be used to benefit humans.

*W*hat makes design interesting is how it can be used with technology to fulfill needs. Knowing both the design process and the considerations that go into design is very important — especially if you want to put the knowledge gained in Chapters 2 and 3 to use. There are many fields in which the use of design is the key to success. By now, you probably need some concrete examples of the design process in action in everyday life. Two dynamic areas in which design is used every day are flight and sound.

Aerospace Technology

*T*oday, intercontinental travel by airplane takes only hours, and it has become commonplace to read about astronauts travelling in space. It is difficult to believe that we have been able to fly for less than 100 years. Flying fulfilled the need to travel faster than by sea and by land.

Early inventors used the bird and the kite as models to build flying machines. People constructed wing-like devices in attempts to learn to fly, hanging onto kite frames and gliding down hillsides and steep cliffs. The first powered flight was made by the Wright Brothers in 1903, over a distance that was less than the length of a modern airliner.

Hot-air balloons

Glider

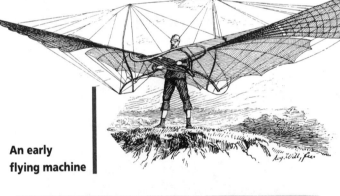

An early flying machine

Passenger jet

The Wright brothers' first flight, 1903

Since then, the development of airplanes has happened at an incredible rate. The shape and performance of airplanes have changed considerably as a result of **aerospace** technology and new materials, but the principles of flight have remained much the same. We can still learn much about flight by observing kites and paper airplanes.

Making a Kite

Imagine that as part of your class work your teacher has decided that each student should make something that flies. To approach this problem, you decide to use the design process outlined in Chapter 2. The following outline shows how this problem might be solved and some of the thoughts you might have during each step.

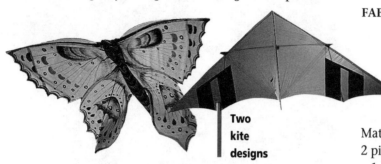

Two kite designs

Need: To create an object that will fly in the air while it is controlled from the ground.

Brainstorming: (possible ideas)
- Create a hot-air balloon.
- Create a model 747 airplane.
- Create a diamond-shaped kite.
- Create a small rocket.
- Create a boomerang.

Design Brief: To create a diamond-shaped kite with lightweight frame and material.

Researching: I could do the following: Discover what makes kites fly. Talk to the science teacher. Check science books and encyclopedias in the library. Consider the best shape and colour, the most suitable materials, and the most effective method of fabrication. Read about kite making and the development and recreational uses of kites.

I should consider these questions: Will my kite be able to fly? Will it be strong enough to withstand strong winds? Do I have enough time? Is my material adequate?

I should draw up a decision-making chart to help me decide which design factors will be important. I will make a few sketches — at least five — with various shapes and measurements.

Planning: I will narrow down the design choices and make a final selection. I should make an actual-size model. I should include a bill of materials. I will use pine and string for the frame and strong paper for the covering. I will also need glue and measuring and cutting tools. A chart will help me to plan the fabrication steps.

I should consider any important safety factors that may arise during fabrication.

Fabrication: I will set up tools in the order shown on my chart. I will construct the kite using the exact measurements of my model, and follow the procedures that I have recorded on my chart.

FABRICATION STEPS FOR THE KITE

Using the following information and sketches, I will be able to produce a classic diamond-shaped kite that will perform well in light to moderate winds.

Materials Needed:
2 pieces of clear pine, 1 cm x 1 cm x 1 m and
  1 cm x 1 cm x 80 cm
strong string
resin glue
sheet of paper (e.g., kraft paper, tissue paper, newsprint)
glue
6 strips of cloth, each 30 cm long

1. Create a **frame** using the two pieces of clear pine. Place the centre of the **spar** (80 cm crosspiece) at a right angle to and 25 cm from the top of the 1 m **spine**. Bind the two spars together with the string and apply a thin coat of resin glue.

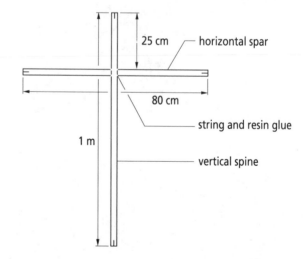

2. After securing the frame, cut notches approximately 1 cm deep in the ends of the spar and the spine to accommodate the framing line. As the **framing line** holds the main frame members in position, it is important to check the frame for squareness and balance before the string is tightened and glued into position.

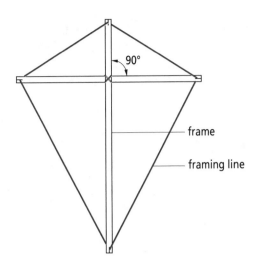

frame

framing line

90°

6. Standing with your back to the wind, holding the flight line, have an assistant hold the kite about 25 m away. The assistant then releases the kite — it will climb as it is forced upward by the wind. Slowly let out the flight line until the kite floats on the undisturbed wind high above the trees.

Framing line and paper attached to horizontal spar

Bridle attached to the main spar through a hole in the paper

Top bridle attachment ¹/₂ distance to top spar

¹/₂

paper

frame

bridle line

¹/₃

Bottom bridle attachment ¹/₃ distance from bottom to horizontal spar

flight line

tail

3. Make a **covering** by laying the sheet of paper on a flat surface and placing the frame on top of it. Trace the pattern, leaving 3 cm of extra paper on the outside of the framing line. Apply glue to the paper on the inside of the framing line and fold over the extra paper.

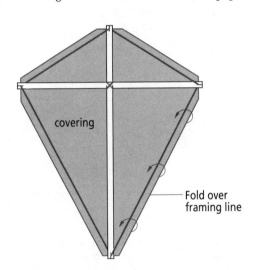

covering

Fold over framing line

4. Glue squares of paper to the covering where the **bridle** points attach. (The bridle is the piece of string that is attached to the kite to hold it in a position that will cause the wind to push up on it.)

5. Tie the **tail** to the base of the spine. The tail, made from knotted strips of cloth tied to a piece of string, will provide a small amount of pulling motion or drag. This drag is needed to stabilize the kite to keep it in the correct position. A good guideline for the tail's length is seven times the length of the spine. The drag factor can be increased or decreased by lengthening or shortening the tail. Recheck all joints and attachment points and tie on the flight line, which should be about 40–50 m long. The kite is now ready for a test flight.

Evaluation: When the kite is completed, give it a series of test flights. In the evaluation, determine whether I have a good design, whether my kite is able to fly in light to moderate wind conditions, and what modifications or improvements are required. Ask myself honestly, "Did I fulfill my need?"

I will also make a presentation to the class and the teacher. In order that others may benefit from my research, development, planning, and testing, I will make a presentation of the beginning stages of the project, sketches, drafts, problems encountered, order of operations, written descriptions, tests, and any changes made to the final kite.

Airplanes

In many ways, the principles that allow kites to fly are the same principles that allow airplanes to fly. The shape of an airplane is dictated by its purpose. If the plane is to glide, the wings are long but not wide. If the plane is to fly long distances at high speed, a long, tapered shape is best. There are many parts in an airplane, and they allow it to fly and make difficult manoeuvres. The diagram below will help you locate each part.

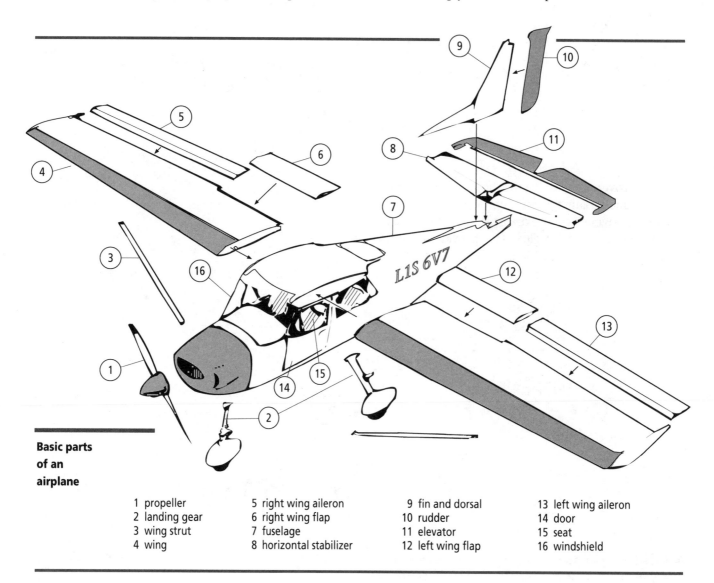

**Basic parts
of an
airplane**

1 propeller	5 right wing aileron	9 fin and dorsal	13 left wing aileron
2 landing gear	6 right wing flap	10 rudder	14 door
3 wing strut	7 fuselage	11 elevator	15 seat
4 wing	8 horizontal stabilizer	12 left wing flap	16 windshield

There are four forces that act upon an aircraft and cause it to fly. They are thrust, lift, drag, and weight.

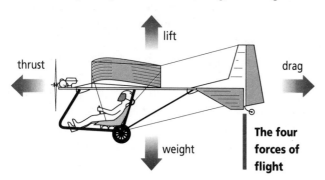

**The four
forces of
flight**

Thrust is the force that is created by the spinning propeller or escaping gases that draw or push the aircraft forward.

Lift is the upward force of air pressure against the bottom of the wings. As the plane moves forward, the pressure increases under the wings until it is greater than the mass of the aircraft. Liftoff then occurs, and the aircraft begins to fly.

Drag is the resistance of the air to the motion of the aircraft.

Weight is the downward force of gravity acting upon the mass of the aircraft.

To achieve flight, an aircraft has to develop enough *thrust* to overcome *drag,* and generate enough *lift* to overcome the *weight* of the aircraft. All four forces have to be in balance to maintain flight.

**How a wing
generates lift**

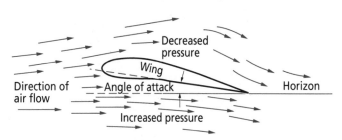

Hold a letter-size sheet of paper in your hand, and let it fall to the ground. You will notice that it arcs from side to side until gravity pulls it down to Earth. Now do this again. You should observe that the sheet of paper will not fly the same way twice.

The challenge is to do something to the paper that will give you the ability to control its flight. In your notebook, sketch some ideas on how the sheet of paper could be changed to make it move through the air in a way that you could predict.

Paper Airplanes

There are three basic types of paper airplanes — the dart, the wing glider, and the conventional airplane. Each type has its own characteristics and abilities, but the one that best teaches the parts of an airplane and demonstrates flight is the wing glider. Suppose that your teacher has asked you to make something that demonstrates the basic principles of flight, using only a piece of paper. Using the following instructions and diagrams, you can produce a wing glider that will demonstrate these principles.

FABRICATION STEPS FOR THE WING GLIDER

1. Hold a letter-size piece of paper horizontally (a piece of notebook paper with the wider edge at the top).

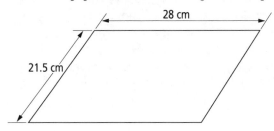

2. Starting at the bottom, make a fold about 1 cm deep. Make four more folds, each 1 cm deep.

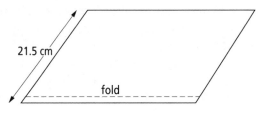

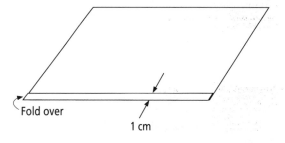

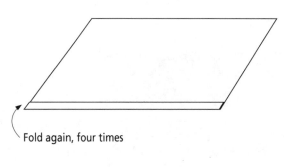

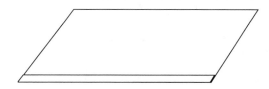

3. Fold the paper in half so that the folded edge is on the inside.

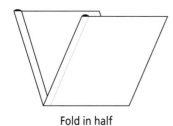

Fold in half

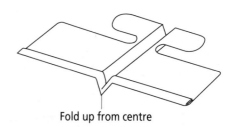

Fold up from centre

4. With the wings folded together, draw the shape of the airplane on the outside.

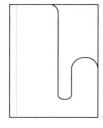

Draw and cut out

5. Cut out the shape you have drawn. Open the wings and tail.

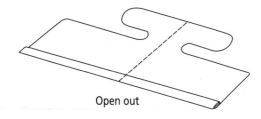

Open out

6. Create a body by folding the wings and tail about 2 cm up from the centre fold.

7. Make tail fins by bending the ends of the tail downward. Make wing fins by bending the ends of the wings upward.

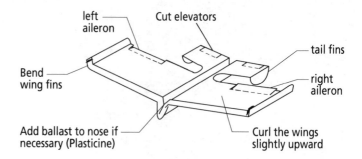

left aileron — Cut elevators — tail fins — right aileron — Bend wing fins — Add ballast to nose if necessary (Plasticine) — Curl the wings slightly upward

8. Cut elevators on the tail to control the plane's flight. Cut ailerons and curl the aileron on one wing slightly upward and the aileron on the other wing slightly downward

9. Experiment with adding ballast (Plasticine) to the airplane's nose.

Evaluation: Toss the glider nose-forward into the air. Watch its flight carefully. Make notes on what you see.

How far did the glider fly? How fast? Did it do what you expected? Did it fulfill your need? Answering these questions will help in evaluating your glider. You should also prepare to make a presentation to your teacher and class.

See for Yourself

Notice how changing the design slightly can change how an object performs. Change the wing glider design by changing the folds and cuts any way you want. Then test the plane and evaluate how it performs. Does it perform better or worse than the original design?

In your notebook, sketch the changes to your design. Beside each sketch, describe the change in the plane's performance.

Design and Flight

Aviators call the three basic manoeuvres made by aircraft the three "axes of rotation." They are roll, pitch, and yaw. **Roll** is the action of turning the length of the plane to the left or right. **Pitch** is the action of suddenly turning upward or diving downward. **Yaw** is the action of revolving around a central point, as in a flat spin.

Roll, pitch, and yaw can be controlled by adding movable surfaces to the aircraft. These new surfaces redirect the flow of air and cause the aircraft to move differently, but also in a controlled way. These adjustable areas are called **control surfaces**.

Roll can be increased or reduced by adjusting ailerons, which are movable sections near the ends of the wings. *Pitch* can be increased or decreased by using elevators, which are movable parts of the horizontal stabilizer. The tail fin reduces the amount of *yaw* and keeps the plane flying on a straight course. The rudder causes the plane to turn in either direction, much the same as a rudder turns a boat. In all cases, the aircraft is controlled by the way the *control surface* redirects the air through which the aircraft is passing.

Modern Airplanes and Spacecraft

The paper airplane, the glider, the propeller, jet-engined craft, and the rocket all work on the same principles of flight. With new technology, the design and fabrication of airplanes have undergone many changes, but the basic principle of *thrust* still applies. The propeller pushes a large volume of air backward at slow speed. The result is a *reaction*, or *thrust*, in the forward direction, much the way escaping air forces a balloon to move.

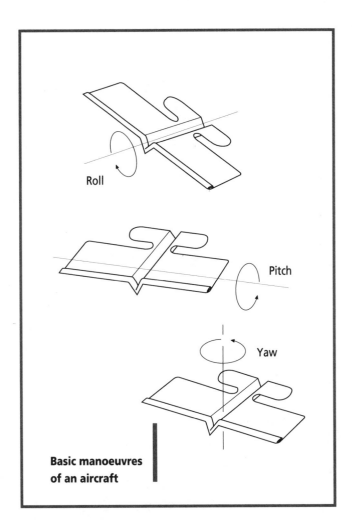

Basic manoeuvres of an aircraft

Roll

Pitch

Yaw

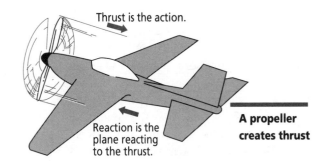

Thrust is the action.

Reaction is the plane reacting to the thrust.

A propeller creates thrust

See for Yourself

The English scientist Isaac Newton said that "for every action there is an opposite and equal reaction." To understand what this means, inflate a balloon. Release the open end, allowing the air to escape. In which direction does the balloon travel? In small groups, discuss ways in which the balloon's movement can be controlled. Test your ideas.

Construct a "guided balloon" system as shown in the diagram below. Notice what happens when you release the clothespin from the opening of the balloon. List the parts of this design that make it work. Use this list in your group discussion.

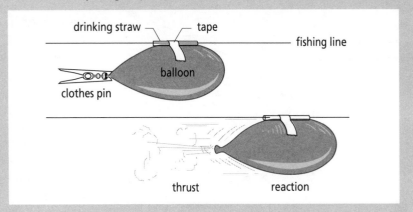

drinking straw — tape
fishing line
balloon
clothes pin
thrust   reaction

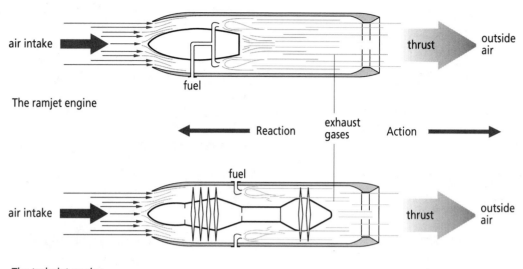

air intake → The ramjet engine · fuel · thrust · outside air

← Reaction   exhaust gases   Action →

fuel · air intake → The turbojet engine · thrust · outside air

It is not the blast of hot gases out of the exhaust of a jet engine that provides the driving force, but rather the reaction to this blast of hot gases. The jet engine sucks in air at the front and releases it at the back. The escaping gases push against the outside air. The entire plane reacts to this pressure by moving in the opposite direction. The harder the push, the faster the plane moves. Unlike the propeller, the jet engine moves a relatively small amount of air at a relatively high speed.

The rocket also uses thrust to propel spacecraft beyond the confines of Earth's gravity into space. Gases are set off by igniting chemicals stored in a tube. As rockets have to be very powerful, there must be sufficient liquid oxygen and fuel reserves to provide the thrust for the liftoff. These fuels mix and burn in a burning chamber. The hot, expanding gases that blast from the tail of the rocket provide the action. The reaction is the force that moves the rocket upward.

The ramjet and the turbojet are different types of jet engines.

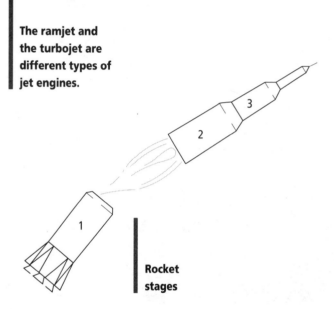

Rocket stages

To make rockets travel farther and faster, the basic design was changed. The rocket was built in sections, or stages, which added power to the rocket's flight. A three-stage rocket has three rockets built one on top of the other. The first stage is for liftoff; this stage is the biggest because it has the most work to do. The second and third stages propel the rocket farther into space. The third stage holds the astronauts and their scientific instruments and carries them toward their goal in space.

The early space missions were very expensive, as few pieces of the rockets were reusable. Today's rockets propel reusable space shuttles that glide back to Earth and land on a runway. The rocket boosters can also be saved and reused in other missions. The new design thus saves time and money.

How a rocket works

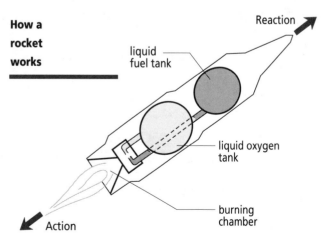

Reaction · liquid fuel tank · liquid oxygen tank · burning chamber · Action

A space shuttle, on the launching pad and in space. This type of spacecraft is a spinoff of the airplane.

Space programs have numerous spinoff effects on technology. These spinoffs happen in the same way you might change an idea slightly in a brainstorming session. For example, satellite technology brought about the development of solar energy cells. These cells were designed to change light into usable electrical energy. Although solar energy cells have been used in communications satellites and space capsules for a long time, they now have many other spinoff uses. For example, pocket calculators, miniature radios, and marine buoys all use solar energy cells.

It is important to realize that all these products resulted from each improvement in aerospace design. It is unlikely that they would exist today if it were not for innovation in aerospace technology.

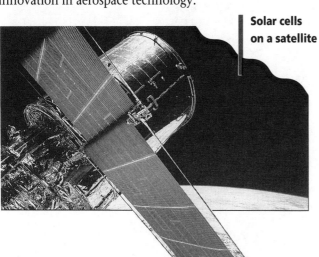

Solar cells on a satellite

Career Profile

Alexander Graham Bell and the Aerial Experiment Association

Most people associate Alexander Graham Bell with the invention of the telephone. However, he was also one of the world's first successful designers of aircraft. One of Bell's long-time interests was human flight. At first, he flew kites of all shapes. He was interested in how the best combination of shape and material could produce the most lift in a "heavier-than-air craft."

Bell had the help of five young men with a sense of adventure: Frederick Baldwin, Thomas Selfridge, J.A.D. McCurdy, Arthur McCurdy, and Glenn Curtiss. Bell and his helpers created the Aerial Experiment Association (AEA) at Baddeck Bay, Cape Breton Island, in 1907. Their goal was to design and fabricate a flying machine to rival and beat the Wright Brothers in their domination of this exciting field.

Left to right: Glenn Curtiss, Frederick Baldwin, Alexander Graham Bell, Thomas Selfridge, and J.A.D. McCurdy.

On July 4, 1908, Curtiss flew the *June Bug* to become the first person to fly a heavier-than-air flying machine for a distance of 1 km under test conditions. For weather reasons, these first flights were held in the United States. But in the winter of 1909, the AEA's Silver Dart flew the first recorded flight in Canada.

With little financial backing, the AEA could no longer support itself, and so it disbanded on March 31, 1909.

QUESTIONS AND EXERCISES

1. What combination of forces for achieving flight interested Bell?
2. Using the design process described in Chapter 2, show how Bell might have designed an airplane that was modelled after kites.

Sound

Design and technology are important not only to science but to music as well. Designing and fabricating musical instruments will help you to understand how sound is created.

Sound is a sensation of particles moving in waves. Most natural sounds are complex variations of many waves.

Sound can be measured in waves.

Oscillation occurs when particles move from a position of rest to another position and then back to their original position.

tuning fork

air

Source ——→ Medium ——→ Receptor

ear

How sound is created and heard

A **medium** is a solid, liquid, or gas through which sound waves can travel. Sound travels at different speeds when passing through different media. It moves faster through solids and slower through gases.

Amplitude describes the maximum distance that a wave particle travels in one *oscillation*. The higher the amplitude, the louder the sound.

Vibration is a rapid movement back and forth.

Frequency is the number of vibrations in a second; it is usually read in cycles per second.

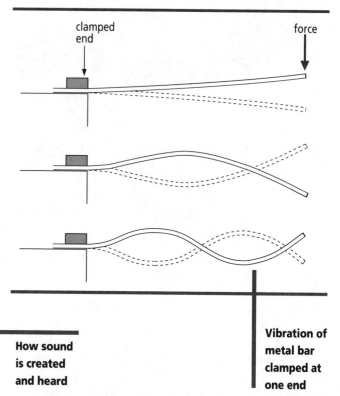

clamped end

force

Vibration of metal bar clamped at one end

See for Yourself

Have two classmates hold on to both ends of a rope. One person holds the rope steady, while the other person moves his or her end up and down quickly. Observe the waves created. Switch roles so that each person has a chance to observe the waves. Now discuss as a group how you think sound waves may act in the same way.

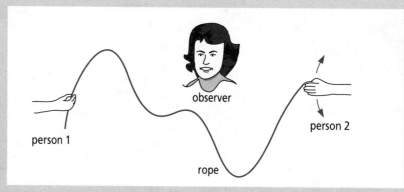

person 1

observer

person 2

rope

See for Yourself

In order to understand how vibrations create sound, tie one end of a piece of string to a coat hanger and the other end to your right forefinger. Do the same with the left forefinger. Then put your fingers carefully into your ears. Hit the hanger against something. When the hanger vibrates, it causes the string to vibrate. The vibrations then travel through the string and your fingers and into your ears. The sound that you hear will be a particular pitch. In order to vary the sound, change the shape of the hanger. Hanging metal objects from the hanger will also change the sound. Experiment with some objects, and compare results with a classmate.

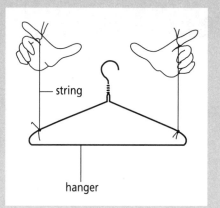

Resonance is a reinforcing and prolonging of sound by the vibration of objects. It enhances sound when it is amplified by a container. For example, a drum resonates when you hit its head because the shell of the drum resonates.

Pitch refers to how high or low a note sounds. The greater the frequency of the vibration, the higher the pitch of the sound.

A **node** is the original position of the sound. It is the point at which there is no vibration, and it is the wave or vibration at rest.

An **antinode** marks the point of the wave at its peak. It is the point at which there is the most vibration.

N = Node
A = Antinode

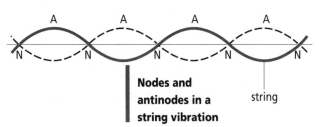

Nodes and antinodes in a string vibration

string

Noise is sound without tone, beat, or rhythm. It is disorganized sound — a complex jumble of oscillations that have no recognizable pattern.

This **decibel (dB) table compares some common sounds and shows how they rank in potential harm to hearing.**

Sound Levels and Human Response		
Common Sounds	**Noise Level (dB)**	**Effect**
Jet engine (near)	140	
Shotgun firing Jet takeoff (30–60 m)	130	Threshold of pain (about 125)
Thunderclap (near) Discoteque	120	Threshold of sensation
Power saw, pneumatic drill Rock music band	110	Regular exposure of more than 1 min risks permanent hearing loss
Garbage truck	100	No more than 15 min unprotected exposure recommended
Subway, motorcycle Lawnmower	90	Very annoying 85—level at which hearing damage (8 h) begins
Average city traffic noise	80	Annoying, interferes with conversation
Vacuum cleaner, Hair dryer Inside a car	70	Intrusive, interferes with telephone use
Normal conversation	60	
Quiet office Air conditioner	50	Comfortable
Refrigerator humming Living room, bedroom	40	
Whisper Broadcasting studio	30	Very quiet
Rustling leaves	20	
Normal breathing	10	Just audible
	0	Threshold of normal hearing

In order to understand oscillation, tape a pencil to the end of a ruler so the pencil is parallel to the ruler. Tape a piece of paper to a board. Hold one end of the ruler firmly on the table and hold the paper to the tip of the pencil. Strike the ruler so that it vibrates; the vibrations of the ruler will be drawn by the pencil onto the paper. The line drawn will show the oscillations, amplitude, and frequency. Be sure to paste the paper into your notebook as a reminder of what sound "looks" like.

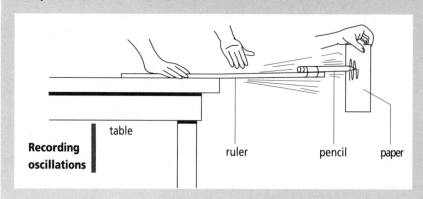

Recording oscillations

table | ruler | pencil | paper

Musical Instruments

When you hit an object, a vibration occurs. This vibration causes sound. An object is resonant when it is capable of vibrating in a regular pattern. A loose string is not resonant, but a tight string is.

A monochord is a musical instrument that has a single tight string stretched between two supports. When you move the bridge (see the diagram), you can change the pitch of the sound.

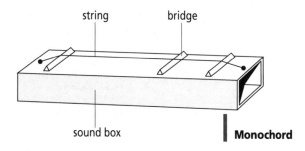

string | bridge

sound box | **Monochord**

The type, thickness, and length of the material used to build a musical instrument will make a difference in the number of oscillations that occur. For example, in order to tune a marimba, the wooden bars have to sit on the nodes at either end of the bar. There will be many nodes and antinodes created when a bar is struck. If the bar sits on an antinode, the pitch of the sound created when the bar is hit will sound out of tune.

Another musical instrument that makes a sound when struck is a chime. Chimes can be made of wood, bamboo, keys, nails, rocks, or glass, and are usually hung from a frame. Aluminum and brass tubes give pleasing sounds.

In order to hang metal tubes to make chimes, you must drill a hole through one end of each tube. The hole must be drilled directly on the node in order to resonate.

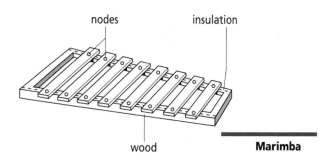

nodes | insulation

wood | **Marimba**

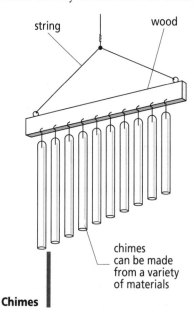

string | wood

chimes can be made from a variety of materials

Chimes

One way to discover where to drill the hole is to hold the tube between your thumb and index finger approximately 7 cm from the end, and let the other end swing freely. Then strike the free end with a mallet. Try this method several times, each time slightly moving the hand that is holding the tube. When the tube resonates loudly after you have hit it with the mallet, you are holding the tube on the node. Mark this spot and drill it exactly on the node.

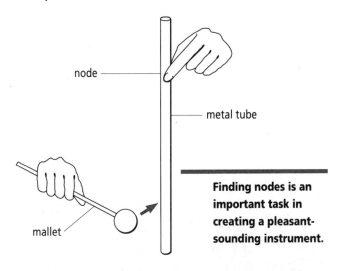

Finding nodes is an important task in creating a pleasant-sounding instrument.

Making a Musical Instrument

Suppose you are asked by your teacher to design and fabricate a musical instrument to show how vibrations create sound. To approach this problem, you decide to use the design process described in Chapter 2. The following outline shows how this problem might be solved and some of the thoughts you might have during each step.

Need: To create a simple musical instrument to show how vibrations can create sound.

Brainstorming: (possible ideas)
• Create a metal triangle.
• Create a slap stick.
• Create a thunder sheet.
• Create a banjo.
• Create a piano.

Design Brief: To create a thunder sheet.

Researching: I could do the following: Talk to the music teacher. Look at existing instruments. Check music books. Make some sketches with various shapes and measurements. Consider the best shape and colour, the most suitable materials, and the most effective method of fabrication.

I should consider these questions: How loud should my thunder sheet sound? Will it show its vibrations? Do I have enough time? Are my materials adequate?

I should draw up a decision-making chart to help me decide which design factors will be important.

Planning: I will narrow down the choices and make a final selection. I should include a bill of materials. A chart will help me plan the fabrication steps. I should consider any important safety factors that may arise during fabrication.

Fabrication: Using the following instructions and diagrams, I will be able to produce a thunder sheet that will sound very loud and will wobble enough to show vibrations.

FABRICATION STEPS FOR THE THUNDER SHEET
Materials Needed:
wood (broom handle)
sheet metal
screws

1. Cut metal with hacksaw.
2. File and sand edges.
3. Cut two lengths of wood for handles.
4. Drill holes in handles for screws.
5. Cut handles in half, lengthwise.
6. Using the holes in the handles as guides, mark where to drill on the metal.
7. Drill holes in metal on ends.
8. Attach wooden handles to ends of metal with screws.

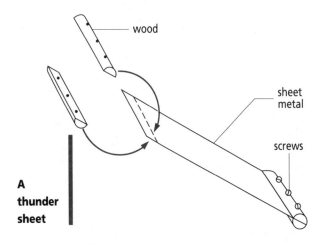

A thunder sheet

Evaluation: Did my thunder sheet match my expectations? Did it fulfill the need? Was it loud enough? Did it show vibrations clearly? What does my teacher think? What should be included in my presentation to the teacher and the class?

Career Profile

Kathy Browning

"In my childhood, I used to create forts with blankets and tables indoors or wood and grass outdoors. Empty shoe boxes became houses for Flintstone characters. I would divide up spaces in the shoe boxes and stare in the windows, projecting my body into the tiny spaces," says Kathy Browning. This early involvement with creating different spaces and considering how bodies move through those spaces encouraged Browning's creative and career endeavours.

Since the age of four, Browning has been involved in performance and music. She also began writing at a young age and showed a proficiency in art. At the age of twelve, she started teaching art, and she organized creative arts programs for the city of Thunder Bay at the age of nineteen.

Browning's grandfather was a carpenter, and she also enjoyed the challenges of woodworking. "While working on my Bachelor of Fine Arts at the University of Manitoba, I built wooden wing images in the architecture workshop that I would then paint." Painting, printmaking, design, sculpture, and photography were her other interests.

While completing her Master of Fine Arts at York University, Browning carved large wooden sculptures, developing her three-dimensional sense. She also began to use discarded materials in her art. Slate from church roofs, shoemaker forms, and metal air vents were transformed into sculptures and shown in galleries.

Later, Browning worked as an industrial designer for a major retail chain, creating three-dimensional sales displays. "I had to design and fabricate structures in multiples. I sometimes made thousands of the same product for the displays. Logos, colour, structural strength, and the transportability of a product to the site are all important aspects of designing

displays. The design process with its creative development, research, organization, and fabrication steps helped me as an industrial designer. This design process is a life skill that can be used in everything you create." Browning had to work quickly and without making mistakes, because she wasn't given much time from when the work order came in to when the display had to be installed in the store.

Browning also creates acoustic musical instruments that have to look and sound good. Browning's musical instruments have been in a variety of art shows in Canada. "One way to check the sound quality of an instrument is by performing live with others." For nine years Browning wrote, performed, and created costumes, props, sets, videos, and cassettes.

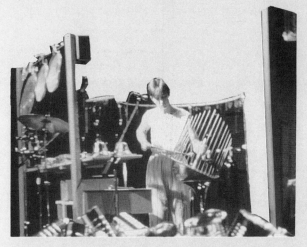

Browning also has a Bachelor of Education and has taught courses in design and technology and the arts from the elementary to the university levels. She is currently completing her Ph.D. in education at the University of Toronto.

Browning is often asked how she can do so many different activities. "I love to create! Although it can be hard work, I have a willingness to acquire the skills to fabricate my ideas. Creating your own designs is very rewarding."

QUESTIONS AND EXERCISES

1. List the types of spaces that Kathy Browning has created from childhood to the present.
2. List the ways in which Kathy Browning's education has assisted her career.
3. Sketch five possible musical instruments. Some of these instruments might be a combination of two instruments, such as a keyboard on a clarinet.

Making a Guitar

There are many simple instruments that you can make. All of them are **acoustic** instruments — that is, they do not need an amplifier in order to be heard. Instruments such as electric guitars need to be connected to an AC outlet and amplified in order to be heard.

An electric guitar is difficult to make, because there are many details and techniques to be learned in order to create this highly specialized instrument. The best way to learn is to create simple acoustic instruments and gain insight into how sound is produced before attempting to create more difficult instruments.

Suppose your school is holding a "lip sync" contest that you and your friends want to enter. Your group decides that you need guitars as part of your act. Below is the design process for making a model guitar.

Need: To build a guitar to be used in a lip sync contest.

Brainstorming: (possible ideas)
- Create a guitar with two necks.
- Create a guitar with one neck.
- Create a guitar out of plastic.
- Create a guitar out of wood.
- Create a hollow guitar.
- Create a solid guitar.

Design Brief: To develop a guitar for a lip sync contest.

Researching: Our group of three will research different shapes of guitars. Then each group member will sketch five different guitar designs, giving the group fifteen designs to choose from. Each group member will decide which is his or her best design and be prepared to present it to the group. The best design will be sketched and a decision-making chart will be drawn up. The group will then hold a conference with our teacher to discuss the selected design and the decision-making chart. This will help to confirm whether we have chosen the best design. A draft of the final design will be made. All group members will make the same design.

Planning: Each group member can help the others through the various stages of design and building. A chart of fabrication steps and a bill of materials should be written. Safety factors that might arise during fabrication will be considered.

Fabrication: We will use basswood for the body of the guitar because this wood is easy to shape. Take two pieces of basswood, each 2 cm thick, and glue edge to edge. Use a paper template to trace the pattern onto the wood, so that all the guitars are the same shape. Make the relief cuts and then cut out the guitar body on the band saw. Make the guitar neck from one piece of wood that is cut to shape. The fret board (top side of the neck) remains flat, while the underside of the neck is rounded. Secure the neck to the body with a lap joint, glue, and screws. The length of the guitars will vary from 60 cm to 85 cm.

Once each guitar is sanded to a smooth finish, it is ready to be painted. A lacquer finish will show the grain of the wood. Using a wood burner to burn in designs might also work. We may also add plastic or knobs for effect. We can decide on certain colours that are in keeping with our band. We may tape out sections to give a striped effect, or paint swirls.

Real guitar strings or string will be added to make the guitars look more real. Guitar straps, long neckties, or string can be used to hang the guitars around our necks.

FABRICATION STEPS FOR THE GUITAR
The following information and sketches will allow the production of guitars for a lip sync contest.

1. After final approval from the teacher for one of the sketches, draw a template of the body of the guitar on paper and cut it out.

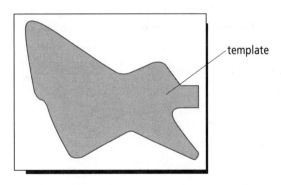

template

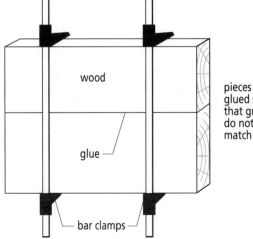

wood

glue

bar clamps

pieces are glued so that grains do not match

Glue and clamp the two pieces of basswood edge to edge for the body of the guitar. Bar clamps should be balanced and near ends. The grain of the pieces of wood should go in opposite directions to avoid warping. (See the diagram on page 53.)

3. Use the template to trace the design onto the wood. Make the relief cuts using the crosscut saw.

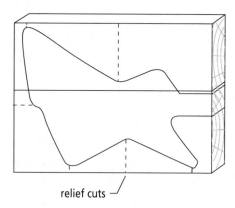

relief cuts

4. Cut out the guitar body on the band saw. File and sand the edges.
5. Draw the guitar neck on the wood. Make the relief cuts with the crosscut saw.

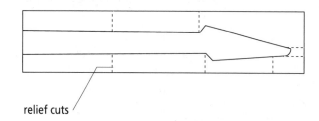

relief cuts

6. Cut a lap joint for the guitar neck using a back saw and file.

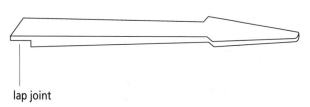

lap joint

7. File and sand the edges of the guitar neck.
8. Drill holes to attach the guitar neck to the guitar body. Use screws to attach the pieces.

9. Put finish on guitar. Add plastic, strings, and knobs for effect.

Evaluation: Ask ourselves whether the guitar fills the original need. Use the guitar as part of the final presentation and evaluation to the teacher and the class. Present our group's notes, sketches, drafts, bill of materials, and charts.

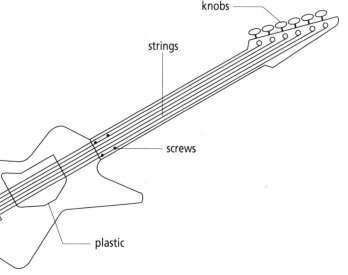

The final guitar design completed

Making musical instruments brings together an understanding of science and music. Our scientific knowledge and the use of design and technology to create music make our world a more enjoyable place in which to live.

knobs

strings

screws

plastic

Points for Review

- The principles of flight can be learned by observing kites and paper airplanes.
- Four forces that act upon an aircraft and cause it to fly are thrust, lift, drag, and weight.
- Three basic manoeuvres made by aircraft are roll, pitch, and yaw, which can be controlled by adding movable surfaces to the aircraft.
- The basic principles of flight apply to gliders, propellers, jet-engined craft, and rockets.
- Sound is a sensation of particles moving in waves.
- Oscillation occurs when particles move from a position of rest to another position and then back to their original position.

- Amplitude is the maximum distance that a wave particle travels in one oscillation. The higher the amplitude, the louder the sound.
- Frequency is the number of vibrations per second.
- Pitch refers to how high or low a note sounds.
- Resonance is a reinforcing and prolonging of sound by the vibration of objects.
- A node is the original position of the sound or the point at which there is no vibration. An antinode marks the point of a sound wave at its peak.
- Noise is disorganized sound without tone, beat, or rhythm.

Terms to Remember

acoustic	frame	oscillation	spar
aerospace	framing line	pitch (in aerospace technology)	spine
amplitude	frequency		tail
antinode	lift	pitch (in sound)	thrust
bridle	medium	resonance	vibration
control surfaces	node	roll	weight
covering	noise	sound	yaw
drag			

Applying Your Knowledge

1. Make a sketch to explain how a kite is able to stay afloat in the air.
2. List the scientific principles that are at work when an airplane maintains straight and level flight.
3. Discuss in a group the following statement: "A passenger jet has a mass of several hundred tonnes. Although it is so heavy, it can fly."
4. List the control surfaces on an aircraft that will cause
   a. turning,   c. climb,
   b. pitch,   d. roll.
5. The airplane has been a very important invention for the transport of people and goods. Research and list the different types of airplanes, based on their use. In a group, discuss how design differs in each type of airplane.
6. Explain, using diagrams if necessary, the scientific principles that cause a rocket to propel a spacecraft.
7. Imagine that you have been given an assignment to write a history of rockets: their early beginnings, their development to the present day, and their importance to the future. List seven sources of information that would help you to write this report. Some of these sources should be in the classroom and some should be outside the classroom. Be as specific as possible.
8. Describe in one sentence what a node is.
9. Sketch five different designs of a musical instrument that you would like to make. Number your designs and complete a decision-making chart (discussed in Chapter 2) to help you select the best design.
10. With the approval of your teacher, do a survey within your school of types and levels of noise. Walk around your school making a list of the sounds, where they are located, and the intensity of each sound. Arrange the list in order from loud to quiet. Share your findings in a group discussion.

Part 1 Projects

The following projects are ready for you to undertake. You have been given the materials and the fabrication steps. The aim of these projects is to give you the confidence to be able to fabricate your designs.

Making a Meat Tray Glider

*T*his project demonstrates form and function and uses surfaces that react to the air.

Materials
styrofoam sheet, less than 1 mm thick, or meat tray
cardboard
paper clip

You can obtain styrofoam sheeting thinner than 1 mm from packaging companies. However, meat trays from the grocery store are excellent, economical alternatives. The rounded edges of the trays make good leading edges for the wings, and the flat surfaces work well for the fuselage and the tail. The remaining pieces with curved edges provide dynamic horizontal stabilizers with elevators. These elevators can be adjusted to control straight and level flight.

Your teacher will provide you with an X-acto® knife and show you how to use it.

© *Caution: Be careful with this tool, since it is very sharp and could cause serious injury.*

Your teacher will also supply you with the other materials.

Fabrication

1. Plan the most economical use of the styrofoam so that all the parts fit on one tray. Make cardboard patterns for the parts.

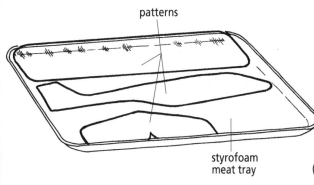

patterns

styrofoam
meat tray

2. Place the front edge of the wing pattern on the curved part of the tray, and trace the shape.

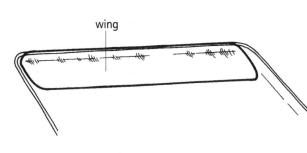

wing

3. Place the fuselage pattern on the longest part of the remaining piece of tray, and trace the shape.

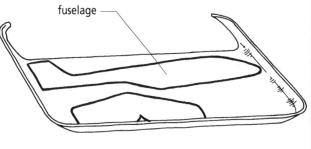

fuselage

(cont'd)

4. Fit the horizontal stabilizer in the part of the tray that is left, and trace the shape.

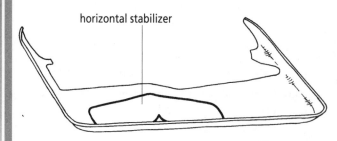

horizontal stabilizer

5. Place the meat tray on a stiff piece of cardboard and carefully cut out the parts with an X-acto® knife.

 Caution: *Handle this tool carefully.*

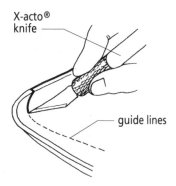

X-acto® knife

guide lines

6. With the X-acto® knife, make a horizontal slot slightly wider than the width of the wing, two-thirds of the way between the tail and the nose on the fuselage. This slot will accept the wings. (Curve the front of the slot downward to allow for the rounded part of the wing.)

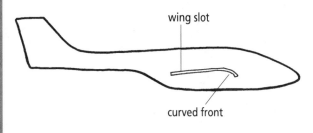

wing slot

curved front

7. With the X-acto® knife, cut a slot near the base of the tail to accept the horizontal stabilizers.

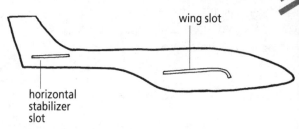

wing slot

horizontal stabilizer slot

8. Assemble the parts of the glider. Use a small paper clip to add weight to the nose, and adjust the elevators up and down to achieve straight and level flight.

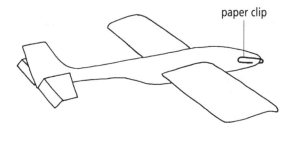

paper clip

9. Larger gliders can be made using additional trays or sheets of 2.5 cm styrofoam.

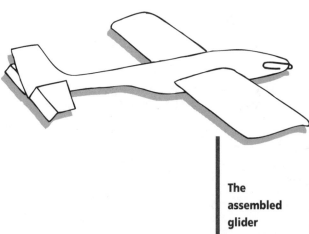

The assembled glider

Making a Light Source

*T*his is an excellent group project, for the group members can develop many possible designs. Each student sketches a minimum of five possible designs. If there are three people in a group then you have fifteen designs from which to choose.

This project is not difficult to make, and it goes with any style of furniture, particularly modern.

Your teacher will supply you with the materials. Choose your own colour of plastic for the lamp shade. A lighter colour will allow a brighter glow than black plastic. It is best to use opaque plastic so that you do not see the light bulb.

Materials

paper
light bulb
light socket
2.5 cm length of threaded rod
lamp cord
cord switch
plug
wood, 2.5 cm x 30.5 cm x 30.5 cm (minimum)
    (1" x 1' x 1')
sheet plastic, 3 mm thick x 25 cm x 76 cm
    (1/8" x 10" x 30")
wood finish of your choice

Fabrication

1. Make a template for the lamp base out of paper. Place this on the wood and trace the pattern. The sides of the triangle should be 30.5 cm long.

Make centre dot to mark drilling location.

30.5 cm
30.5 cm
30.5 cm

Paper template for lamp base

2. Cut out the triangles. File and sand the edges of one triangle. Drill a hole in the centre of the wood. The size of the hole should match the OD of the threaded rod.

3. Cut a 2.5 cm length of threaded rod. This will be threaded into the hole later. Use a round file to make a notch under the base of the lamp from the hole to one side of the triangle. This will hold the lamp cord so that the lamp doesn't rock on the cord.

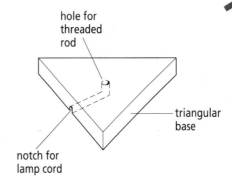

hole for threaded rod

triangular base

notch for lamp cord

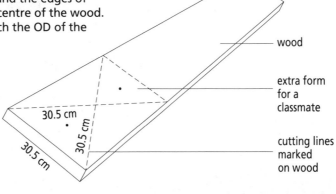

wood

extra form for a classmate

cutting lines marked on wood

30.5 cm
30.5 cm
30.5 cm

(cont'd)

4. Cut the plastic so that you have a piece that measures 25 cm wide x 76 cm long. There are many ways to cut plastic. A fine blade in a band saw works well. When filing the edges of plastic, always make sure that you secure the piece with wood, because the plastic will crack if it is moved back and forth too abruptly. The edges should be sanded and buffed.

5. Draw lines on the plastic to divide it into three equal sections. The paper on the bottom side of the plastic should be removed so it will not catch fire. Use a strip heater to heat the plastic until it becomes pliable. Bend the plastic on the lines to make a triangular shade. A jig can be used to help get the correct angle on the bends, or you can put the plastic on your wooden base and bend it to the proper angle, holding it until the plastic cools. It is best to bend the plastic slightly farther than you want, because it tends to spring back a bit when it cools. Making a triangular shade with only two bends is much easier than making a square shade with three bends.

6. Apply the wood finish to the base.

7. Screw the piece of threaded rod into place, then screw the bottom of the socket onto the rod. Splice the lamp cord and strip it. Put it through the hole in the lamp base, and thread it through the rod and socket base. Connect the wires to the positive and negative on/off switch. Assemble the remaining parts of the light socket. The plug can be put on at home. Remove the paper from the outside of the plastic, and place the lamp shade on the base.

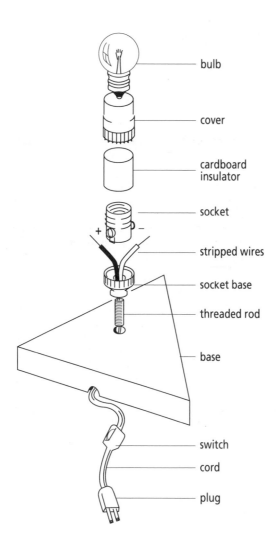

- bulb
- cover
- cardboard insulator
- socket
- stripped wires
- socket base
- threaded rod
- base
- switch
- cord
- plug

The finished wood and plastic lamp

Making a Wind Tunnel

*T*he fabrication of a small-scale wind tunnel will allow you to test the aerodynamics of projects that you design and fabricate, such as scale-model automobiles or airplanes.

Materials

16 1-L milk cartons
masking tape

Fabrication

Tape the cartons together in four rows of four.

To use, place the wind tunnel in front of a variable-speed electric fan. Use book ends or other supports to keep the wind tunnel in place. The fan produces the moving air, and the wind tunnel channels the air into an even flow. Hang the model to be tested from a support — this support could be another project for you to design.

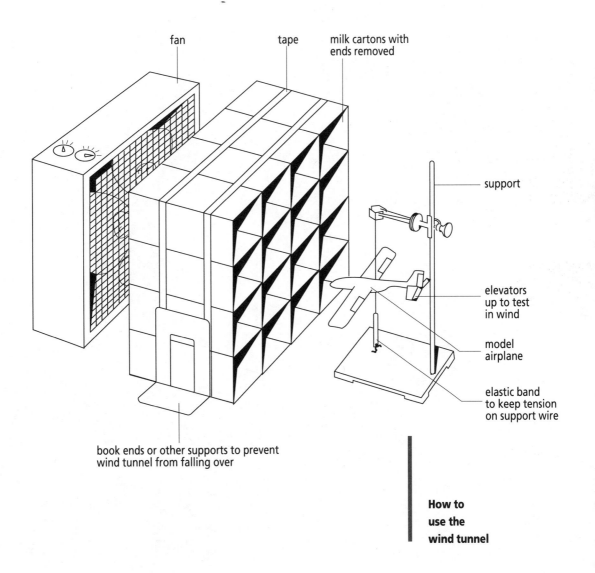

fan

tape

milk cartons with ends removed

support

elevators up to test in wind

model airplane

elastic band to keep tension on support wire

book ends or other supports to prevent wind tunnel from falling over

How to use the wind tunnel

Design Solutions —
Starting from Scratch

*T*he following are some situations that require design solutions. If you feel that you are ready to meet the challenge of starting from scratch, read the situations carefully and then use the design process described in Chapter 2 to find solutions. Be sure to consult your teacher after each step, especially before you begin fabrication.

1. After conducting a survey of four classes, it is discovered that most students have a problem with too little locker space. They find that their lockers are too small and that things get cluttered easily. Use the design process to solve this problem.

2. Most homes need a suitable container for sorting cans, paper, and bottles for recycling and materials for composting. Use the design process to design and fabricate a container that will fit under the kitchen counter and that has separate compartments for sorting materials.

3. With increased urbanization, more wildlife habitats are at risk of being destroyed. Use the design process to design and fabricate a feeder for your yard or park that will accommodate birds, squirrels, or rabbits.

Part TWO

Research and Development

Chapter 5

STRUCTURES

What You Will Discover

After completing this chapter, you should be able to:

- Understand how engineered objects and structures are designed to function.
- Understand why things and people stand up.
- Identify the forces that act upon structures.
- Know the importance of shape and materials in the design of bridges, towers, and houses.

*A*s you have already learned from your work in Part One, designers must know many things to design products that fulfill needs. They must understand technology, because designers use technology as a resource to create possible solutions for needs.

Some types of engineers do work that is similar to the design process. They understand basic principles of science and mathematics; they test and develop new materials; and they do extensive research into the design and location of a structure.

A **structure** is anything composed of parts arranged together. Before any large structure (in the form of buildings, bridges, or towers) is fabricated, engineers must calculate which force will act on each part of the structure and choose or design the part that will resist that force. **Force** is a push from any direction. The material used to build the structure must also be able to withstand the greatest amount of wear caused by water, wind, sun, temperature, and pollution.

When engineers are designing a bridge, for example, they carefully analyze the forces acting on the structure and the projected loads it will carry. The bridge must be strong enough to support not only its own mass but also any mass that may be placed on it in the future. In parts of the bridge where the only force will be tension, for instance, fairly thin rods and cables can be used.

Engineers have to use materials that will be able to withstand environmental forces and pollution. This is why structural engineers use plastics for I beams when creating large structures such as highrise buildings. Modern plastics can withstand tremendous loads and environmental wear, or

The Eiffel Tower in Paris, France, is a world-famous structure.

The CN Tower in Toronto, Canada, is another of the world's great structures.

deterioration. The melting point of these plastics is high, which means that they can withstand extreme heat caused by fire.

Basically, a structure is only as strong as the framework that holds it together. If the framework is strong and made out of the proper materials, it will last a long time and be more cost-effective.

Forces

*T*here are many forces acting upon structures. Force can be defined as a push or pull in a given direction. For example, our mass placed on a chair exerts a downward force. When a force acts on an object, that object will deflect away from the force in proportion to the thickness of the object.

Forces that act when sitting in a chair

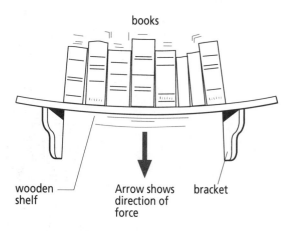

Arrows show forces and their direction

A shelf will bend under the mass of books if it is suspended between two brackets. This mass is a force. The amount the shelf will bend depends on the thickness of the shelf and what it is made of. Every material has different characteristics that determine whether it is useful for a required application.

Forces that push on a structure are called **loads**. There are two types of loads. Anything that is placed on a structure is called a **dynamic load**. When you walk across a bridge, for example, you are a dynamic load. The books on a shelf are a dynamic load. Anything that is part of the structure is called a **static load**. All of the nuts and bolts, rivets, steel, plastics, cement, or anything else used to fabricate a bridge are static loads. The wood, glue, nails, and varnish used to make a bookshelf are static loads.

When a board is placed between two supports, it will sag if a sufficient load is applied to it. If the supports are far apart, the board may sag a little even when there is no dynamic load on it. The mass of the materials that make up the structure create a static load on it. So, any structure must be not only strong enough to support its own mass but must also be strong enough to support any dynamic load that may be placed on it for any length of time.

Compression

Compression is a *pressing* force. When compression forces become too great, buckling, or giving way under the strain, occurs. Some materials avoid buckling more effectively than others. Sometimes the centre of a long bridge will be supported by a **pier** made of brick, con-

crete, or stone. The mass of the bridge compresses the pier. The pier must be built in such a way that repetitive compression will not push it out of place.

Often a long bridge will need concrete piers to act against compression.

However, not all compression force is downward. Water freezing around a pier causes horizontal compression that, together with the undermining effect of the moving water, may cause the bridge to collapse under the strain. Compression from wind and from loads such as cars or people also affects the bridge.

See for Yourself

When a material does not hold up under compression, it buckles. To understand buckling, stretch a rope using both hands. Bring your hands together. As your hands come toward each other, the rope is unable to provide any resistance and so folds up.

Tape a book to a piece of uncooked spaghetti. Lift the book using the spaghetti. Place the book on a table, and use the spaghetti to try to push the book. Why is the spaghetti able to lift the book but not push it? After trying this, you will be better able to understand why a chain is able to tow a car but is not able to push it. Use your observations to take part in a class discussion on compression.

Tension

Tension is a *pulling* force caused by stretching a material. In the case of tension forces, no buckling occurs. Long, thin materials such as metal rods, ropes, and steel cables are suitable for resisting tension provided they are not overstretched.

Many things are held in place by tension forces, such as pictures hanging on a wall and lighting fixtures hang-

ing from a ceiling. Machines depend on tension to operate: weights that drive the gears of some clocks hang from chains, and emergency brakes on cars are activated by pedals attached to cables that pull the brakes. Sounds can be produced from steel wires in tension: the strings of a guitar can be tightened until they produce the desired sounds.

If we could look inside a material that has a load applied to it, we would see that there are two forces at work: compression and tension. The top of the material is pressed downward and the bottom is pulled apart.

Guitar strings use tension to produce sound.

Cuckoo clocks run on the tension forces from weights (pine-cones) hanging from chains that drive the gears.

See for Yourself

Make several small cuts in the top and bottom of a 2 cm x 2 cm strip of styrofoam insulation provided by your teacher. Rest the foam strip on two blocks of wood. With your finger, press down lightly on the centre of the foam. Observe the cuts in the top when you press down. What did they do? What happens to the cuts on the bottom? The pressed top shows the effects of compression, and the spread-open bottom shows the effects of tension. Record your testing methods, and discuss with a classmate whether any material would show the same effects of compression and tension as the foam.

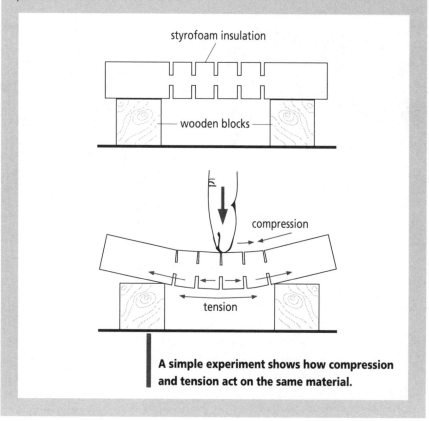

A simple experiment shows how compression and tension act on the same material.

Cast-iron beams are usually made thicker on the tension face than on the compression face, because cast iron cannot withstand as much tension. A feather, as light as it is, is carefully designed by nature to prevent it from failing when placed under a load. The quill of the feather is a box-like structure with a foam-like core. This core is reinforced with lengthwise ribs on the tension side to provide strength for the feather with each downward beat of the wing in flight. The compression side has no ribs.

The box-like structure of a feather quill

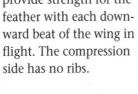

compression face

tension face

Rib structure gives strength to the tension side of the feather.

Materials and Function

*B*efore designers or engineers select the materials for building the various parts of a structure, they must know which material is best suited for each purpose. Each material withstands forces in different ways, and so must be tested for every possible force to which the structure will be exposed. (For a more detailed discussion of the importance of materials, see Chapter 8.)

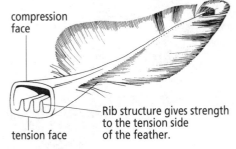

Shape and design work together in these structures.

Some materials that are suitable for one application may not be suitable for other applications. Some materials, for example, are able to withstand compression but fail under the strain of tension.

Materials and Structures

Examining structures used for transportation over history helps us to recognize how the choice of materials changes with each new need.

Bridges are typical structures needed for transportation. Stone was one of the first materials used successfully to build bridges. Stone is able to tolerate compression but is quite brittle under tension. Most materials that are hard and brittle exhibit these qualities.

Early stone workers invented the arch when they joined two pillars with wedge-shaped stones. As other stones were placed on top, the wedges were pressed tightly together. This design provided the strength that was needed to overcome the inadequacies of the material.

A basic arch

The Colosseum in Rome shows the use of arches.

An arch built by ancient Romans. Do you think it was designed for function or decoration?

The standard arch is still used for viaducts.

As time went on, the arch was refined and was later used in the fabrication of aqueducts, bridges, and cathedrals.

The scarcity of stone made it more and more difficult to find enough material for long bridges. Stone was also too heavy to transport great distances. Eventually, a human-made form of stone — concrete — was developed as a substitute for natural stone. Both stone and concrete had the same problem, however, when used as the main material for bridges. Both could withstand much compression, but broke apart quite easily under tension.

Another material that was commonly used for building bridges up to the nineteenth century was wood, which was more plentiful than stone. Wood could withstand tension and compression, but it could not stand up to weathering, chewing insects, or fire.

Iron replaced wood in bridge-building when nineteenth-century engineers learned how to pour melted iron into forms for fabrication purposes. The iron could be produced quite cheaply, and builders could take advantage of its great strength.

As rail transport became more popular, train engines became faster and heavier, and they carried larger loads. Cast-iron girders that performed well under compression soon deteriorated from the tension placed on the undersides of the girders as they flexed from the increased loads.

Iron became a popular material for bridges in the nineteenth century.

See for Yourself

Rest a toothpick on two small blocks of wood, and then apply pressure with your finger. As you press harder and increase the load, the toothpick will begin to break. The top part of the toothpick is compressed while the bottom is pulled apart by tension. Stand another toothpick upright and apply pressure to the top with your finger. Be careful not to drive it into your finger like a sliver. What happens? In groups of three, debate the following statement: "Almost any material can be used to build a structure."

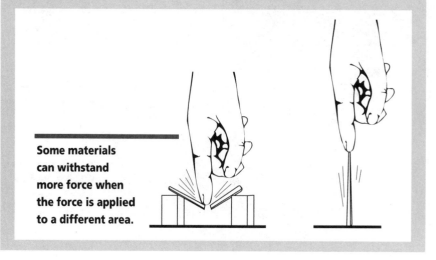

Some materials can withstand more force when the force is applied to a different area.

Steam locomotives rumbled down the tracks carrying freight and passengers at faster and faster speeds. Day after day, trains pounded across the iron bridges. And then, after standing for many years, the bridges began to collapse.

Engineers investigating the problem discovered that the bridges had failed from **fatigue**. This condition occurs when metal is subjected to continuous application and removal of a force. If you bend a paper clip back and forth several times, the wire will weaken and finally break. This is due to fatigue. As a result of fatigue, many iron railway bridges built during the 1800s had to be crossed at slow speeds. The use of iron in bridge fabrication soon gave way to reinforced steel, a stronger and more durable material.

Today, modern **suspension bridges** are built from a wide range of materials that can withstand maximum compression and tension forces. The main materials used are steel cables and concrete.

The Lions Gate Bridge in Vancouver, British Columbia, is a perfect example of a suspension bridge.

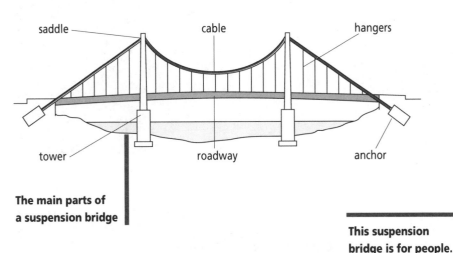

saddle   cable   hangers

tower   roadway   anchor

The main parts of a suspension bridge

This suspension bridge is for people.

Suspension bridges span long distances by using high-tensile steel wire. This material is very strong and is capable of supporting great mass. The three main supporting parts are the towers, cables, and anchors.

The concrete towers are the first parts of a suspension bridge to be built. Cables containing many thousands of strands of high-tensile steel wire are passed over the tops of the towers and set on huge pulleys called saddles. If the cables were anchored at the top, they would pull the towers over as mass was applied to the centre of the bridge. The saddles roll and allow the cables to move, yet the cables are secured by the anchors embedded deep within the sides of the riverbank.

When all of the main supporting parts are completed and in place, the level roadway must be hung from the main cables. Large sections of roadway are hung from cables called hangers, and then the roadway is permanently fastened in place.

Using Triangular Shapes

Often the shape of the material used will determine the shape of a structure. For instance, stone was cut into triangular shapes to form arches, because it was recognized that the triangle had qualities that could withstand forces that other shapes could not. A structure made in the shape of a **triangle** will resist any change of shape when forces are applied to it.

Shelves supported by triangular brackets that hang on a wall demonstrate the principles of the triangle. When a load is applied to this type of shelf, the force is downward and is transferred toward the wall. The brackets press against the wall, preventing the shelf from collapsing.

Career Profile

Allan Baker

"Sometimes it pays to have fun, doing what you love to do," says Allan Baker, educator and designer of special living environments. Throughout his education and career, Baker has returned to the discovery that doing what you enjoy most, while using your skills, can often be the best way to decide on a career.

When he was young, Baker discovered that he had the "aptitudes and interests of a designer." He was fascinated by the performance and appearance of objects. "For instance, if it had wheels and moved, I liked it, changed it, broke it, fixed it (sometimes), and then improved it."

After secondary school, Baker studied industrial design at the Ontario College of Art. Upon graduating he worked in various jobs, always returning to those areas that gave him the greatest satisfaction. "During my career, I have worked as a designer in the broadest sense of the word. My design activities have included graphic, exhibit, industrial, furniture, interior, and architectural design, as well as teaching design education," says Baker. For example, he worked as a technical illustrator for an aircraft manufacturer and as a designer and draftsperson for consulting firms. Some products that he helped design and fabricate were logos, hockey helmets, and snowmobile models. Later, Baker became principal designer and proprietor of his own design consulting firm. It was "exciting, risky, and rewarding," he says, "but I wished I had taken business management courses along the way."

Eventually, Baker became involved in teaching design part-time. This forced him to "re-think why I did things as a designer." The big challenge came when he was asked to help establish a new industrial arts program at Georgian College. At first he had only planned to help for a few years to get the program going, but then he found the change exciting. He became the full-time chairman but later a full-time instructor, "realizing that working with students was what I really enjoyed."

Baker has continued to do design work while a teacher. One of his major projects during a professional leave of absence from teaching was to develop products for independent living. This involved designing alternative housing for senior citizens, whose living accommodations need special consideration.

Baker feels strongly about design education and is committed to its place in Canada: "This country desperately needs creativity and innovation in its secondary industry. When it awakens to this fact, through enlightenment or crisis, it will finally embrace, respect, and foster the Canadian design talent that is available. Until that time comes — and it will — we should continue to develop our professional capabilities internationally while making inroads domestically where we can. Design will play a major role in the future of this country."

QUESTIONS AND EXERCISES

1. What did Allan Baker do when he was young with objects that had wheels and moved?
2. In what design activities has Allan Baker been involved?
3. What are some of the products that Allan Baker helped to design?
4. List a minimum of five products that seniors might need to assist them daily.

Trusses

A triangle will resist moderate pressure. However, when the load becomes too great, the triangle can buckle and fail. Under conditions where the triangle may fail, braces are used to support it. A brace is a support used to strengthen a triangular, square, or round shape. The method of placing the braces forms triangles within the shape. This structure is called a **truss**.

Trusses are usually triangular in shape. A simple truss has no braces. A single post truss is braced vertically from the top to the bottom. A braced single post truss is called

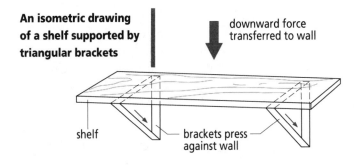

An isometric drawing of a shelf supported by triangular brackets

downward force transferred to wall

shelf

brackets press against wall

a king post truss. Two single post trusses turned so that the bracing appears as a W make a **Warren truss**.

Roofs of buildings are supported by a series of trusses. Locations that receive large amounts of snow require well-supported roofs. An A-frame house is simply a covered, reinforced collection of triangles.

Typical Truss Bridges

Bridges that must sustain heavy loads, such as freight trains, are supported by trusses. These form a framework that gives added strength to the main beams. A bridge that spans a long distance and that uses many trusses or braced frameworks of wood or metal is called a **trestle bridge**.

A cantilever bridge is a truss-like structure that balances on a pier. A long cantilever bridge is a series of truss bridges, each balancing on piers. To construct a cantilever bridge, a large pier must be built in the middle of the area to be spanned. Balancing from each side of the pier and anchored securely to it is the lower part of the truss bridge. Each end of the bridge extends outward in equal stages so as not to overturn it. The final parts are joined together by short truss or trestle bridges.

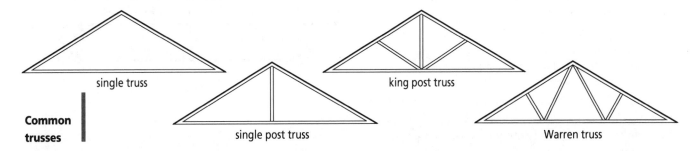

single truss

Common trusses

single post truss

king post truss

Warren truss

See for Yourself

To test the strength of a triangle, cut a strip of cardboard about 10 cm long and fold it in the middle. Shape the cardboard into an inverted V, and place it on a flat surface. If you press on the point of the inverted V, observe that the sides spread outward. When the point is pulled up, the sides come together. To form a base for the triangle, cut a second piece of cardboard about 8 cm long. Fold it upward about 2 cm from each end. Fasten these 2 cm tabs to the first piece of cardboard with paper clips, closing off the open side of the V (see the diagram). Now repeat the pressing and pulling tests. What happens? Make notes on your observations of the V shape.

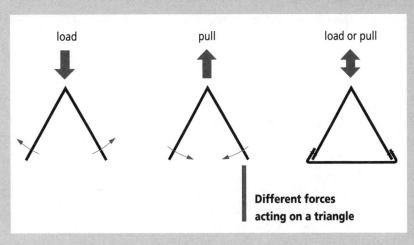

load

pull

load or pull

Different forces acting on a triangle

Note the use of trusses in this cantilever bridge.

I will consider the following questions: Will my bridge be strong enough? How long will it take me to fabricate? What will I use to join the pieces of supporting material?

I should draw up a decision-making chart to help me decide which design factors will be important.

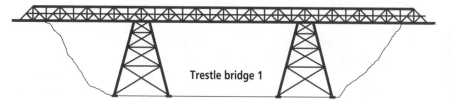

Trestle bridge 1

Making a Common Trestle Bridge

In Chapters 2, 3, and 4 you learned how to use the design process to fulfill needs. When given a certain situation, you should use the design process to help you solve the problem.

Every day, teams of engineers are called on to solve problems. Many times a solution requires the fabrication of a structure. The structure is put through a long design process, including all the stages of finding the need, writing the design brief, brainstorming, researching, planning, fabricating, and evaluating. The situation described below shows how you might go about attempting to fulfill a need by designing a structure.

Suppose that you have a model railroad set that needs something to cover the gap over a small "river." The following describes how you might use the design process and complete the steps in your design and technology class.

Need: To create an object that will allow the train to travel over the "river."

Brainstorming: (possible ideas)
- Create a catapult that will propel the train.
- Create a boat that will carry the train.
- Create a suspension bridge.
- Create a trestle bridge.

Design Brief: To fabricate a trestle bridge of readily available material, strong enough for the train to make repeated trips.

Researching: I could do the following: Borrow books from the library that show typical bridge structures. Consider the materials that are readily available (for more on material choices, see Chapter 8). Determine whether one of my teachers knows more about this topic.

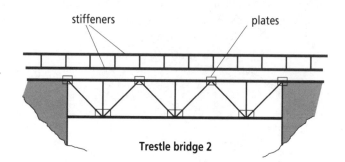

stiffeners   plates

Trestle bridge 2

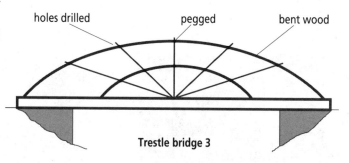

holes drilled   pegged   bent wood

Trestle bridge 3

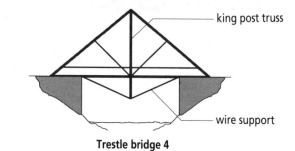

king post truss

wire support

Trestle bridge 4

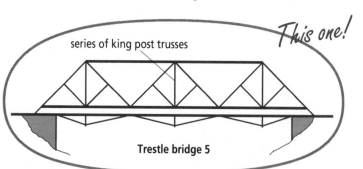

series of king post trusses

This one!

Trestle bridge 5

Planning: I will start by measuring the height of the train, the length of the planned span of the bridge, and the width of the track to be laid over the bridge. Then I will make some sketches of a common trestle bridge. The sketches and other research will help me to make a final selection. My bill of materials should include 1 cm x 0.5 cm strips of pine and a glue gun. A chart will help me to plan the fabrication steps after conferring with the teacher about all my plans and research. Safety factors such as sharp cutting tools and the hot glue gun should be considered before starting.

Fabrication: I will use my final sketches, drafts, and fabrication chart (as okayed by my teacher) to produce the trestle bridge.

FABRICATION STEPS FOR THE TRESTLE BRIDGE

Materials Needed:
pieces of pine, 1 cm x 0.5 cm
hot glue gun

1. Using the same size of wood for all parts of the bridge (1 cm x 0.5 cm, cut to different lengths), begin by making two equal-sized triangles. Brace the triangles to make a king post truss. Make three trusses like this.

2. Sandwich the bases of the trusses between two main beams and similarly join them at the middle and the top. When one side is completed and glued, build a duplicate.

3. Place the two finished sides in an upright position slightly apart from each other, and join them with pieces of equal length along the base to form a roadbed.

4. Support the top with pieces of equal length, and brace the corners for strength.

Rough orthographic drawing of the selected design

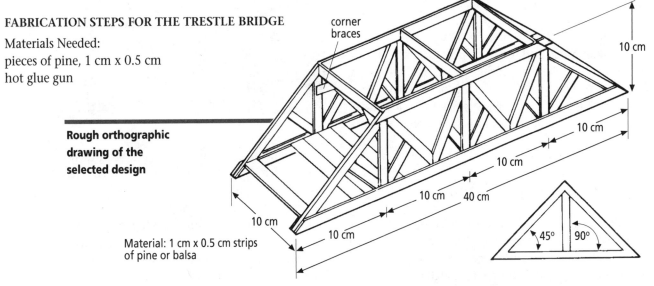

Material: 1 cm x 0.5 cm strips of pine or balsa

Fabrication steps for the trestle bridge

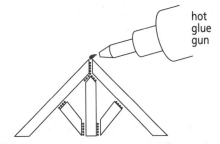

Material: 1 cm x 0.5 cm strips of pine or balsa

Make 2 sides and join together with 10 cm strips to form a roadbed and supports for the top

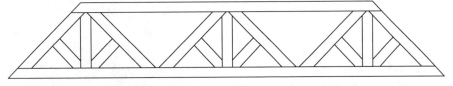

Trusses sandwiched between 2 strips at the top and 2 strips at the base

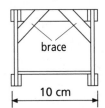

brace

10 cm

Evaluation: When the bridge is completed, run the electric train through it several times. Test the bridge using the method described below. In the evaluation, check to see whether the design brief has been fulfilled. Was the need satisfied? Have all the design factors been fulfilled? Were materials used efficiently or wastefully? What does my teacher think? Prepare for class presentation.

TESTING THE BRIDGE

Weigh the bridge. Test the structure by suspending a pail from its centre. Two tables act as the supports (abutments) for the bridge. Add sand slowly to the pail until the bridge shows signs of failure.

Weigh the pail of sand and weigh the total expected load on the bridge: the track plus the total number of cars on the train. If the pail of sand has a greater mass, the bridge will be strong enough.

You can calculate the efficiency of the bridge (how much dynamic load each gram of static load will support) by dividing the mass of the pail of sand by the mass of the bridge. For example, if the pail of sand has a mass of 1 kg and the bridge has a mass of 5 g, the efficiency is 200 g — every gram of static load will support 200 g of dynamic load.

Shapes of Girders and Beams

Long, flat, rectangular plates of steel are used in the construction of bridges and buildings. These plates are called **girders**. Even when a load is placed on these girders, it is difficult to bend them. However, although they do not bend, they do twist and buckle out of shape. For this reason, flanges and stiffeners are used to provide additional strength. **Flanges** are flat pieces of metal welded or riveted to the long edges of the *plate*. **Stiffeners** are pieces of metal welded or riveted to each side of the *girder*.

Usually two or more girders, parallel to each other, are used in the construction of a bridge. The girders provide a great deal of strength by themselves. In places where they are joined, a pier is built for extra support. Steel braces are laid between the bases of the girders, and a road or railway is built between the girders.

Girders supported by a pier

Steel **beams** are usually smaller than girders and have different proportions. They are often referred to as I beams or engineering beams. On modern highways, short bridges are usually supported by steel beams or reinforced concrete, while long highway or railway bridges are most often made entirely of girders. The mass of girders limits their use to a span of not much more than 35 m unless supported by a pier.

In large commercial buildings, the framework is made exclusively of I beams of different dimensions. The beams' uses range from the framework of the highest skyscraper to the floor joists supporting a house. They are found in the shortest bridge and in the longest railway.

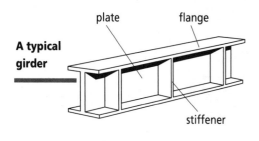

A typical girder

plate   flange   stiffener

Construction of the World Trade Centre in Halifax, Nova Scotia. Notice the many beams used in the framework.

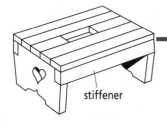

stiffener

The same engineering principles that are used in building bridges apply to building simple stools.

See for Yourself

To see why I beams are stronger than flat beams, try the following. Lay a flat ruler between two wooden blocks. Press the centre of the ruler. It will offer little resistance and bend easily. When the ruler is laid on its edge, it shows remarkable strength against pressure applied to its centre. A flat material has little strength when lying flat and great strength when lying on edge. With a classmate, discuss and record how you think I beams are used in typical buildings.

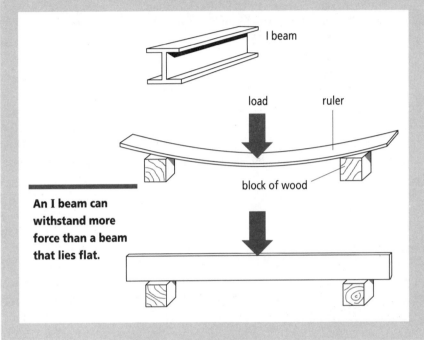

I beam

load   ruler

block of wood

An I beam can withstand more force than a beam that lies flat.

Towers

Throughout the ages, **towers** have been built for many purposes. Engineers are able to build towers to lofty heights by using triangles, squares, and rectangles in their construction. These shapes provide a structure with strength as well as with a strong base that allows the structure to stand without additional support.

Towers are built from stone, brick, wood, or steel. They can stand on eight legs, four legs, or even one leg. Bracing is used to support some towers, but some stand on their own. The Leaning Tower of Pisa in Italy, the Eiffel Tower in France, and the CN Tower in Canada are unique structures and important landmarks in each country. The CN Tower is an architectural feat: it is currently the tallest free-standing tower in the world. Tall, thin towers such as those used for the transmission of radio, television, and other telecommunications signals are able to reach great heights by supporting them with a series of long, strong cables known as **guy wires**. The wires are held in place by large cement anchors.

The Leaning Tower of Pisa. How would you design a solution to stop it from falling over?

The Eiffel Tower, looking upward from the base. Compare this photo with the photo on page 64.

A bell tower, as its name implies, is a high tower that contains a bell. The height allows the sound of the bell to be heard over a great distance. In the past, the bell would warn the residents of a town of some danger, or it would call them to a meeting.

Watch towers were built in the past for defence purposes. Castles had turrets built at the top of watch towers so that guards could monitor the activities of the enemy.

Fire towers are small houses built on top of metal frameworks. They provide a base for forest rangers to look for fires in heavily treed areas.

Water towers are containers that hold water high above the ground. Gravity draws the water down from the tank, creating water pressure at the tap. In some municipalities, water is dispensed to homes by means of such towers.

Transmission towers support equipment that sends and receives signals for television, radio, and other telecommunications systems.

Hydro-electric towers hold wires that carry electricity across long distances.

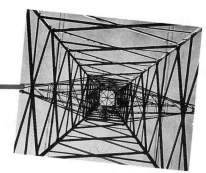

A hydro-electric tower, looking upward from the base. How many shapes can you see?

Modern Housing

One of the most common structures that you can find anywhere is a house. With the increased demand for housing, engineers, architects, contractors, and builders have developed methods and materials to construct houses quickly, efficiently, and economically. In the design of a house you will find all types of materials in a wide range of shapes. This is because of all the different needs that the structure must fulfill.

Once the site and position of a house are selected and the design is drawn, construction begins. A hole is dug below the frostline for the cement **footing**, or base. When the footing has hardened, cement blocks are laid until they reach above the ground level. These provide the **foundation** for the house. Weeping tile is placed around the foundation to provide a place for water to drain. The tile is covered with gravel to provide easy access for runoff water to enter the tile. The wooden **sills** are secured to the top row of blocks with anchor bolts. A girder spans the length of the building to give additional support to the floor beams, or **joists**. Piers are built to support the centre of long girders.

When the parts of the foundation are in place, the floor joists are nailed to the sills and the girder. The floor joists, measuring 5 cm x 25 cm, are spaced 40 cm apart and held in place by a header board to which they are nailed. They form a framework to which the floor is attached. In houses built before 1950, the joists were covered with a **subfloor** made by nailing 15 cm boards diagonally across the floor joists. The modern flooring technique uses sheets of plywood nailed to the floor joists.

With the flooring in place, the **sole plate** is nailed on top of the floor around the perimeter of the house. A modern and quick method of construction is to build a complete wall section on the floor and stand it up when finished. The sole plate can then be nailed to the studs from the bottom and the top plate can be nailed on at floor level.

A water tower. Why do you think it has this shape?

A radio transmission tower. Notice the different shapes that make up the structure.

A hydro-electric tower

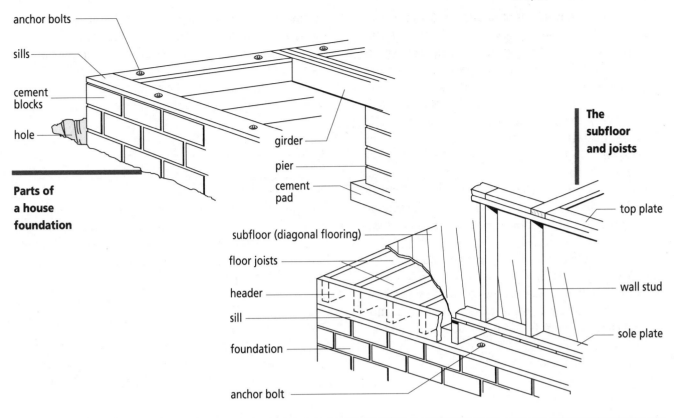

anchor bolts

sills

cement blocks

hole

Parts of a house foundation

girder

pier

cement pad

The subfloor and joists

top plate

subfloor (diagonal flooring)

floor joists

header

sill

foundation

anchor bolt

wall stud

sole plate

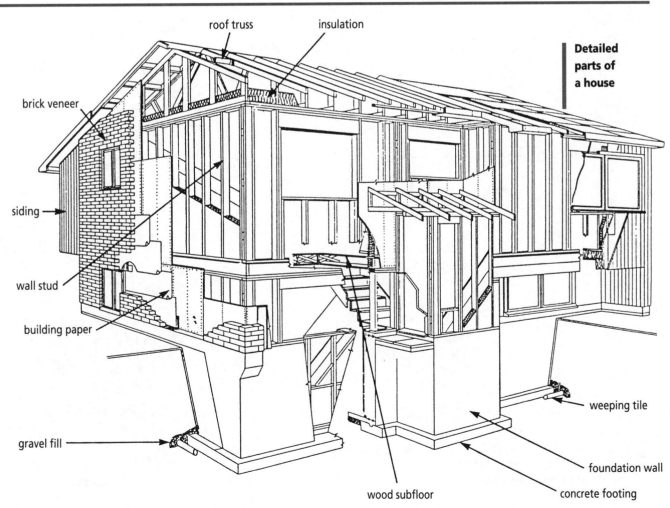

roof truss

insulation

Detailed parts of a house

brick veneer

siding

wall stud

building paper

gravel fill

wood subfloor

weeping tile

foundation wall

concrete footing

Studs are upright pieces of wood spaced 40 cm apart to form the framework of the walls. Windows and doors must have double studding to accommodate the width of the trim. The **trim** is the decorative wood around windows, doorways, and the bottom of the walls.

The framing pieces that make up the roof are called **rafters**. When the rafters are secured in place, the builder adds the **sheathing**, or outer covering of plywood, and nails down a layer of tarred paper to make the structure weatherproof. To protect the roofing, shingles are laid.

The framework of the walls is also protected from the elements by brick or by wooden, plastic, or aluminum **siding**. Brick is still the most popular covering for the exterior walls of houses.

Before the interior walls are insulated and covered, the electrical and plumbing work has to be completed. All work has to follow strict building codes and standards.

In energy-efficient housing, the environmental and natural elements must be taken into consideration.

Deciduous trees lose their leaves in the winter, allowing the Sun to provide additional warmth to the house. In the summer, the leaves provide shade to keep the house cool. Coniferous trees have dense, bushy needles. When these trees are planted on the north side of the house, winter winds and blowing snow are blocked, keeping the house warm.

Any part of the roof that faces south can be fitted with solar-heating panels. Similarly, energy-efficient windows can be installed in the south wall. These measures will save fuel costs in the winter.

Below the frostline, the ground maintains a constant temperature. A house built in the side of a hill or under a mound of earth remains cool in the summer and requires a minimum amount of supplemental heat to achieve comfortable temperatures in the winter.

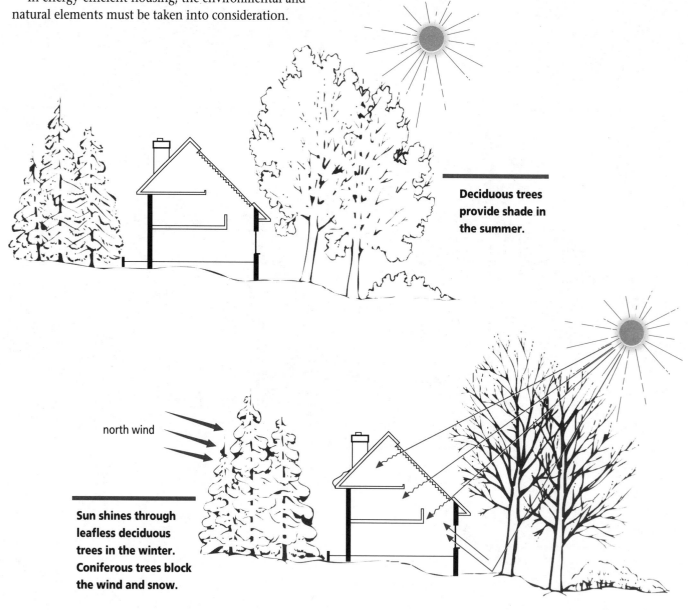

Deciduous trees provide shade in the summer.

north wind

Sun shines through leafless deciduous trees in the winter. Coniferous trees block the wind and snow.

Points for Review

- A structure is anything composed of parts arranged together.
- When engineers design a structure, they have to consider the forces acting on the structure as well as the wear on the materials.
- Forces that push on structures are called dynamic or static loads.
- Compression is a pressing force.
- Tension is a pulling force.
- Fatigue occurs when a material such as metal is subjected to continuous application and removal of a force, leaving it weak.

- A structure made in the shape of a triangle can withstand forces that other shapes cannot.
- Trusses, which are made up of triangles within other, larger shapes, are used in bridge and roof construction to support a great mass.
- Girders are strengthened with flanges and stiffeners that are welded or riveted to the long edges and sides of the girders respectively.
- I beams are smaller than girders and have different proportions. They are typically used in the construction of short bridges and skyscrapers.

Terms to Remember

beams	foundation	siding	suspension bridges
compression	girders	sills	tension
deterioration	guy wires	sole plate	towers
dynamic load	joists	static load	trestle bridge
fatigue	loads	stiffeners	triangle
flanges	pier	structure	trim
footing	rafters	studs	truss
force	sheathing	subfloor	Warren truss

Applying Your Knowledge

1. Explain in a brief paragraph how a material can be exposed to the forces of compression and tension at the same time.
2. Describe the many different forms that environmental factors and pollution can take. How can they affect the deterioration of a structure?
3. The elements of weather that have worn down mountain ranges are still affecting many built structures, such as those from ancient civilizations (for example, the Sphinx in Egypt).
   a. Identify these elements and research how they erode.
   b. Make a chart showing your findings.
   c. Suggest ways that structures could be protected from these elements.

4. Sketch a hydro tower. Show the engineering principles that have been used to overcome forces that would destroy the structure.

5. In a group, discuss the following question and record your ideas: How can builders of physical structures such as bridges, towers, and houses live in harmony with nature rather than against it?

6. Consider the following need: "I need a stool for use at home." Using the design process, design a stool to fulfill this need. Your stool must be designed so that the legs will not spread outward when a load is applied to the seat.

7. What will make it possible for large structures to be assembled in space? Design a space platform to be used for accommodation, relaxation, and refuelling.
    a. Brainstorm five sketches.
    b. Select one sketch as your final choice.
    c. Complete an orthographic or isometric draft of your design.
    d. Write on your draft what needs are being met by the various parts.

8. What is a dynamic load? Explain briefly and give two examples.

9. Consider the following scenario: You are a contractor who has to build a highway bridge within a short period of time. More time on the project will take more money, but less time may result in a less satisfactory structure.

    In a group, discuss the pros and cons of finishing behind, ahead of, and on schedule.

10. Using a well-labelled diagram, explain how a beaver uses engineering techniques to build a dam and thereby stop the flow of water. (HINT: Think of need, materials, and shape.)

11. Why did railroad bridges built in the 1800s later collapse at an average of twenty-five bridges each year?

Chapter 6

ENERGY

What You Will Discover

After completing this chapter, you should be able to:

- Recognize the energy sources around us.
- Indicate new ways of creating energy.
- Understand our dependence on non-renewable energy sources and investigate alternative sources of energy.
- Understand the benefits and drawbacks of energy production.
- Identify energy consumption and ways to reduce it.
- Identify elementary principles of electricity and electronics.

*B*roadly defined, **energy** is the capacity for doing work. It is the force that makes you active. It is also the force that makes things move. When a form of energy is applied to a machine or mechanism, it moves to perform some function. For example, when a motor is energized by electricity, it causes a shaft to turn. As the shaft turns, gears, pulleys, or cranks move. Electricity is only one form of energy.

Designers need to know how to generate energy to fulfill present and future needs. They also need to know about the forms of energy that exist, where energy comes from, and how energy can be applied. Designers need to look at what it takes to create new energy and what is created — besides the new energy — that may be harmful to the environment.

Forms of Energy

*H*ow does energy affect our everyday lives? Though you may experience energy only after it has been converted into various uses, it can be recognized easily, for energy is everywhere. There are several common forms of energy.

Kinetic energy is the energy of an object in motion. We observe kinetic energy in a can when it is kicked, in a hockey puck when it is shot, and in a basketball when it is bounced. The faster an object moves, the more energy it has. The greater the mass of an object, the more kinetic energy it is able to exert.

Potential energy is stored energy that can be converted into other forms of energy. A flashlight battery is a source of stored energy. When dried firewood is burned, the stored energy escapes in the forms of heat and light. When a bowstring is pulled back, it has stored energy and so does the arrow that rests against the string. When the string is released, its potential energy is converted into kinetic energy. The same energy is transferred to the arrow, and sends it into flight.

Gravitational energy is similar to kinetic energy in that it is energy created by motion. However, this type of energy is dictated by gravity — the energy created by a falling object. For example, a falling weight has enough gravitational energy to propel a rock from a catapult.

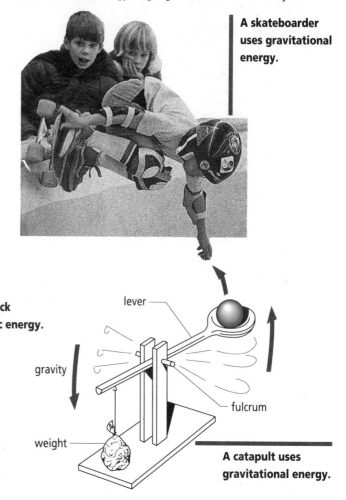

A skateboarder uses gravitational energy.

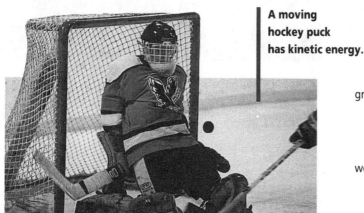

A moving hockey puck has kinetic energy.

lever

gravity

fulcrum

weight

A catapult uses gravitational energy.

See for Yourself

Roll any type of ball down a ramp so that it hits a wall or other barrier at the bottom. Observe the results. Roll a styrofoam ball of equal size down the same ramp. Which ball has more force? Which ball carries more kinetic energy? Sketch these balls in your notebook, and write brief notes on the factors that give each ball its force.

In order to understand gravitational energy, fill a large pitcher with water. Pour the water from a height into a pail. Have a classmate place the edge of a waterwheel (made from a styrofoam cup) in the flow from the pitcher. Gravitational energy causes the water to fall; when it strikes the waterwheel it causes the wheel to turn. Be sure to switch roles with your partner. Predict how changing the height of the "waterfall" will affect the wheel, then test your prediction and observe the effect. Replace the axle with a crank. What kind of work could this machine perform?

Thermal energy is produced when the molecules of a material are heated and then begin to move. The more the molecules are heated, the faster they move and the more energy they create.

Radiant energy is light energy that is sent out from a source. The radiant energy from a light bulb travels as light in straight lines from every part of the outside of the light bulb. The Sun gives out enormous amounts of radiant energy upon which all plant and animal life depends. Our society has yet to fully utilize the Sun's energy. There are many ways we can design our structures to ensure that a maximum of solar energy is used.

Sound energy is produced by vibrating objects. Some objects vibrate more rapidly than others. The vibrations move the air, creating sound waves that we hear as sound. (For a more detailed discussion of sound, see Chapter 4.)

Electrical energy is formed when very small particles called electrons move through a material, creating a current. Electrical energy powers many of our modern machines. This form of energy is discussed in more detail later in this chapter.

Magnetic energy is created by the attraction and repulsion of magnetically charged materials. Magnets have poles, which are two or more points where most of the magnetic flow is concentrated. Opposite poles attract each other. Like poles repel each other. A magnet can be used to do work when it is combined with other parts in a motor.

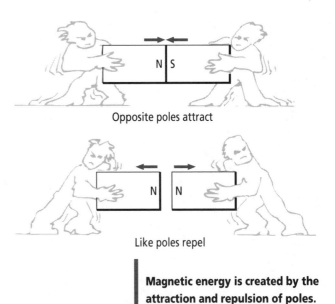

Opposite poles attract

Like poles repel

Magnetic energy is created by the attraction and repulsion of poles.

Chemical energy is released when heat or electric current is introduced to chemical substances. Fuels such as gasoline contain chemical energy, which can be used in engines to power vehicles. The burning liquid fuel in a rocket provides the energy that is required to launch a space vehicle.

All substances are made up of microscopic particles known as atoms. An atom is made up of a central particle, or nucleus, and revolving electrons. **Nuclear energy** is the potential energy that is stored in the nucleus of an atom.

In order to release the energy of the atom, a process of **nuclear fission** must occur. To accomplish nuclear fission, the nucleus of one atom is split. The energy that is released splits the nucleus of other atoms and creates a chain of reactions and collisions. In a nuclear reactor, the heat created by fission is controlled to heat water, which in turn creates steam to power the electrical generators.

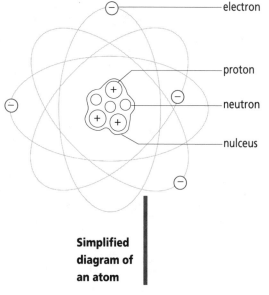

Simplified diagram of an atom

Sources of Energy

*K*nowing the sources of the general forms of energy is very important from a design perspective. Designers have to know how readily available a source of energy is, because this will determine the cost of using the energy. Our use of energy has grown steadily over the years. Industrialized nations, in particular, are highly dependent on energy to meet their needs. This energy dependence has led to increasing costs and decreasing resources. Today, there is an urgency to conserving energy and to seeking new or alternative sources of energy.

Fossil Fuels

Traditional sources of energy have to be found and heavily processed to be usable. Examples are coal, petroleum, and natural gas. These are called **fossil fuels**, because they originate from decayed and compressed plant and animal matter from thousands of years ago. Fossil fuels are among the most widely used sources of energy. Unfortunately, these resources are **non-renewable**. That is, once they are used, they cannot be replaced.

Nuclear fission

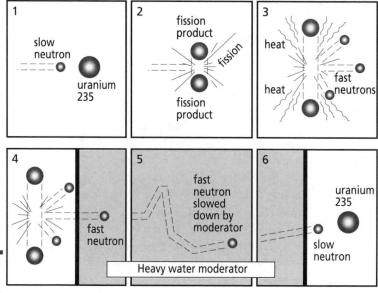

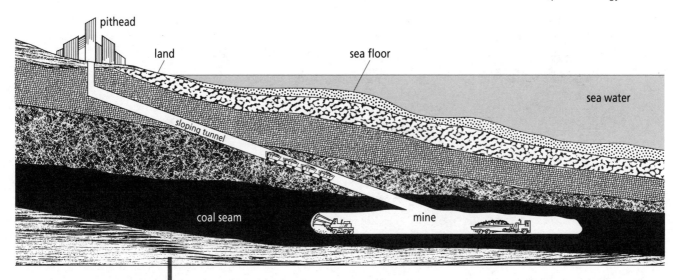

Coal has been of great benefit to industrialized nations, but it has a negative impact on the environment. The burning of coal has been one of the largest contributors to acid rain. Coal also has to be mined from the Earth, and there are numerous dangers to miners, such as cave-ins, flooding, choking gas, inhalation of coal dust, and explosions.

Coal mining requires people and heavy machinery working at great depths.

Petroleum, also called crude oil, is extracted from layers of rock or from the sea floor. It must be refined before it can be used. The process of refining crude oil is called **fractional distillation**. Crude oil is made up of a number of chemicals, each with its own vaporization point. The crude oil is heated, and when each chemical reaches its vaporization point, it turns into vapour and flows into the fractional tower. This tall, steel column is heated so that the top is cooler than the bottom. The vapour enters the heated part and swirls upward, beginning to cool. Each part of the crude oil remains a vapour until it reaches its condensation point and turns into a liquid. Some of the liquids that are produced from this separation are gasoline, kerosene, and diesel fuel.

Recovering petroleum from the Alberta oil sands will provide a new source of fuel for Canada. However, the extraction process releases pollu-

tants into the air and results in hazardous wastes in the form of leftover granular particles.

The transportation of crude oil has also caused concern. Huge ships called supertankers transport crude oil across the seas, but have also caused giant oil spills. These spills have destroyed the plant and animal life on many shorelines around the world. On land, pipelines that travel through remote areas can develop leaks that

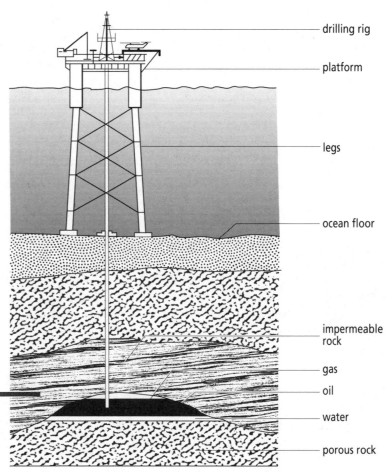

An off-shore oil rig. This is an example of a structure being used to extract energy.

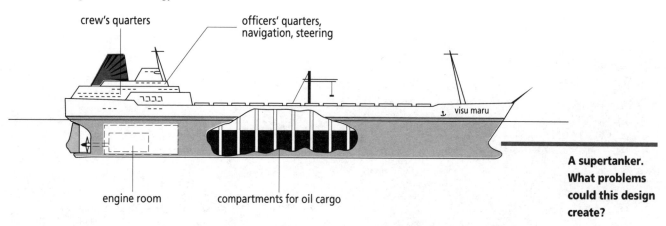

crew's quarters

officers' quarters, navigation, steering

visu maru

engine room

compartments for oil cargo

A supertanker. What problems could this design create?

go undetected for days or weeks, polluting the land. Pipelines can also disrupt the migration routes of animals such as the caribou. The effects on the environment have to be taken into consideration in our quest for new energy sources.

Natural gas is one of the gases that make up crude oil. It is mainly composed of methane, a gas created from decayed organic material. Other components of natural gas are propane, butane, and ethane. Natural gas is often found when drilling oil wells, but it can also occur on its own. Natural gas needs less refining than crude oil.

THE GREENHOUSE EFFECT
Much of the Sun's radiation is reflected back into space by clouds. But the Earth's atmosphere also acts like a greenhouse, trapping some heat. We call this the **greenhouse effect**. When too much heat is trapped, the entire climate system of the Earth is affected.

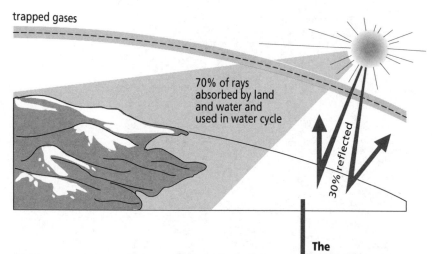

trapped gases

70% of rays absorbed by land and water and used in water cycle

30% reflected

The greenhouse effect

The greenhouse effect is causing global warming and is changing the Earth's weather patterns. We don't yet know all the possible harmful effects that this could have on life, but we can predict some of them. Much of the world's agriculture depends

on a stable weather pattern. When this is endangered, so are the crops and the economy based on these crops. Global warming could also cause major droughts. Crop failure and drought together could cause widespread famine, particularly in less-developed countries. These are only a few of the dangers of the greenhouse effect.

In order to minimize the greenhouse effect, we need to reduce the amount of heat that is trapped by the clouds. Carbon dioxide is the most important of the trapped gases. It acts like a blanket around the Earth, holding in much of the Sun's radiation. The level of carbon dioxide in the atmosphere has been rising steadily since the 1800s, due mainly to the burning of fossil fuels in homes and industries and by vehicles.

Reducing the greenhouse effect will require finding new ways of processing fossil fuels so that harmful gases are not released into the atmosphere. It might also require finding new sources of energy. The non-renewable nature of fossil fuels and the harmful nature of their by-product gases make a strong case for finding other, more beneficial sources of energy.

Sun

The Sun has always been a source of energy. Its energy comes to us every day in its heat, although cloudy days pose a problem if we need to use the Sun as a continuous energy source. This does not mean that we should ignore solar energy as possibly the best alternative to fossil fuels. The Sun is a **renewable** source of energy, because we do not use it up — we are not taking energy away from it.

In order to make the Sun work for us, we have to develop the capabilities for collecting, storing, and converting solar radiation. There are two systems for using solar energy: active and passive.

Hot water is used for a variety of domestic and industrial uses. An active solar water-heating device uses a mechanical pump to move or circulate the water. A passive solar water-heater moves the water by convection currents in the water; the warm water, not being as dense, floats on the cooler water.

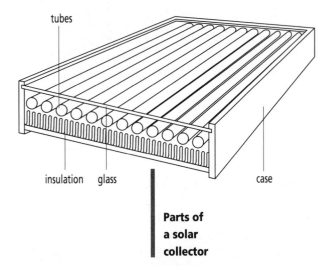

Parts of a solar collector

In an active solar energy system, the solar collector is also used to produce hot air. The Sun's heat is absorbed by black metal tubes. The heat from the tubes is transferred to air inside them. The hot air is blown through pipes by a fan to a rock storage area. The rocks are in an insulated box and are heated by the hot air; they hold the heat until it is needed. The heat is then sent through the building by a fan or convection current.

The passive solar energy system allows heat to flow by natural means. Passive systems are most widely used in houses. An example of the passive system is a thermo-storage wall. This wall is built in a location where the heat from the Sun will be absorbed by the masonry wall. The wall heats up during the day and releases its heat as the temperature drops at night. Masonry walls and a heat storage floor also work on the conduction and radiation principle. **Conduction** is the transfer of heat from one particle to another. **Radiation** is the giving out of rays.

Reflection is another method of increasing heat. The law of reflection states that *the angle of incidence is equal to the angle of reflection.* An example of this principle can be seen in a flat mirror that reflects the light straight back to the source. However, a curved mirror will reflect the light to a focal point. This is the place where the light rays will converge or meet. If the mirror has a dish or

See for Yourself

Make a parabolic cooker using a make-up mirror. Hold the magnifying side of the mirror to the Sun.

Using a piece of paper, find the spot where the light rays converge to the smallest point. At this point, a great deal of heat is produced.

 Caution: *Do not look into the mirror — you could suffer permanent eye damage. The paper at the focal point will burst into flame, so be careful.*

Suspend a piece of meat (try a hot dog) at this focal point. The intensity of the light rays will cook the meat.

Parabolic mirrors are used to cook food (in camp stoves and hot dog cookers) and to boil water for steam (in huge water containers that create enough steam to turn turbines).

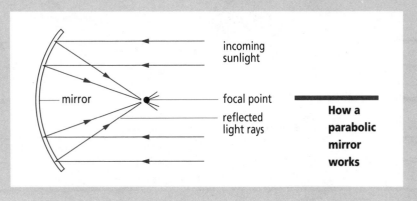

How a parabolic mirror works

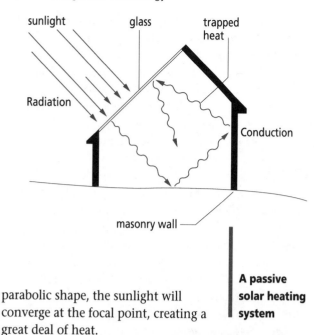

Sailing ships used the force of the wind to explore the world and to carry supplies and people to other nations. Today, sailboats are used mainly for recreational purposes.

A clipper ship runs on wind power.

A passive solar heating system

parabolic shape, the sunlight will converge at the focal point, creating a great deal of heat.

Photovoltaic cells, or solar cells, are thin wafers of silicon that convert sunlight into electrical energy. These cells can convert sunlight to such uses as 12 V of power to charge batteries to run water pumps, lamps, and radios. You have probably seen calculators that run on solar cells. On the ocean, a blanket of solar cells mounted on a boat's cabin roof can charge a battery that operates navigation equipment. The light from channel markers shines all night long because solar cells have stored energy during the day in batteries. In the future, car roofs may be equipped with solar cells to ensure a fully charged battery and frost-free windows as a car sits in a parking lot.

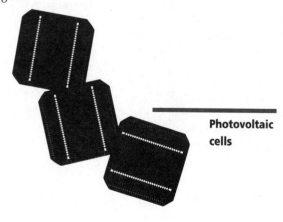

Photovoltaic cells

Windmill blades have been made of different materials over the years: animal hides, heavy cotton, wood, metal, and lightweight aluminum. The shape and number of the blades have also varied, from four large rectangular blades to many curved metal blades on a circular metal frame.

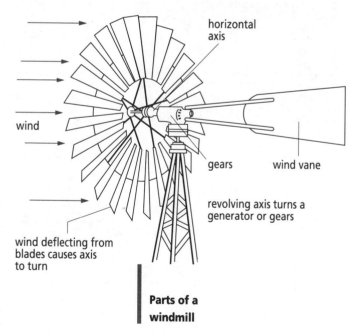

horizontal axis

wind

gears   wind vane

revolving axis turns a generator or gears

wind deflecting from blades causes axis to turn

Parts of a windmill

Wind

Wind energy is a renewable source because it is not used up. At one time, wind was the only natural source of energy that was used to move machines. Wind pushed the large blades of a windmill, causing an axle to turn a millstone or a water pump.

Windmills are now used to generate electricity. Modern windmills, called wind turbines, have few blades but are designed to be highly efficient. The wind-powered generating machines on windy hills look like giant egg-beaters. The initial cost of a wind-powered generating system is high. However, the wind is free and non-polluting.

sunlight   glass   trapped heat

Radiation

Conduction

masonry wall

An old windmill in Belgium

New wind turbines in Belgium. Compare this design to the old windmill. Which do you think is more efficient?

of operating household appliances directly from a wind generator. However, they will work only as long as the wind is blowing. Alternating current is the current in a wall outlet.

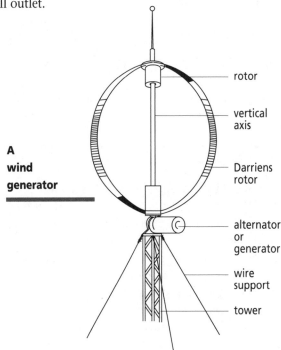

A wind generator

- rotor
- vertical axis
- Darriens rotor
- alternator or generator
- wire support
- tower

One large industrial wind generator is capable of supplying the electrical needs of several homes. Generators that are turned by the wind have an output of direct current, which can be stored in batteries. **Direct current (DC)** is an electric current that always flows in one direction. However, if the energy is used for household appliances, an inverter must be used to change the flow to alternating current. Direct current is the current produced by a battery or solar cell.

An alternator produces electrical energy with an **alternating current (AC)**. This type of electric current constantly reverses direction. Alternating current is capable

Uranium

Uranium is a radioactive material found in most rocks, soils, and rivers. Material that is **radioactive** releases particles that interfere with the natural growth and health of other living things. This is why the careful processing of these materials is so important.

Uranium is extracted from the Earth in much the same manner as coal. A great deal of heat is produced when the atoms of uranium are split in two in the process of nuclear fission. When just one uranium atom splits, it gives up a small amount of heat, but when billions and billions of atoms split, the large amount of heat produced can boil water to make steam to turn generators that produce electricity.

Atoms move at 42 000 km/s. If an atom can be slowed to 3 km/s, it can be steered to collide with other atoms, starting a chain reaction. A nuclear reactor is designed to promote a chain reaction in uranium and to remove the

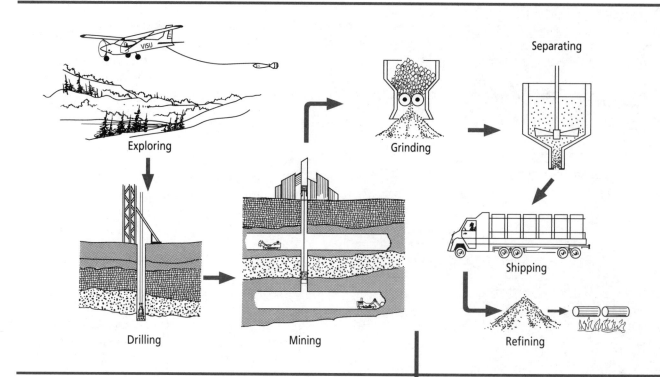

Exploring

Grinding

Separating

Drilling

Mining

Shipping

Refining

heat produced by this reaction. Water circulates between the spaces of the tubes containing the uranium in the reactor and takes away the heat that is produced by the reaction. The resulting hot steam turns a turbine to generate electricity. Several types of reactors exist, but they all work on the same principle. The Canadian reactor is called the CANDU, which stands for **Can**ada, **D**euterium, and **U**ranium.

A major drawback of nuclear energy is that the nuclear reactor produces harmful radioactive wastes. The

Processing uranium

solid waste from fission is still radioactive. It cannot be used again for fuel purposes, so it must be disposed of safely. This is difficult, since any failure in a disposal area could expose the immediate area to terrible long-term risks. If living things are exposed to these products for a long period, they can suffer severe health problems or even die. If the waste falls into the wrong hands, it could be used to make nuclear weapons. Radioactive steam may affect a nearby area if it is released in large quantities, or if it repeatedly affects the same area over a long period of time.

Nuclear reactors are costly to build and maintain. They have a life span of twenty to thirty years, and then they have to be torn down, disposed of, and replaced.

There is no doubt that nuclear energy has been of significant benefit to society. But its safe use must be weighed against the risks

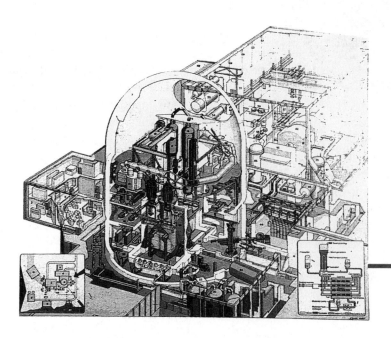

A nuclear power station

in order to evaluate whether the cost is worth it. Nuclear energy offers a perfect example of how designers must look at the pros and cons of everything that goes into a design. By fulfilling one very important need — such as the inexpensive generation of consumer energy — technology may be causing other, significant problems.

Water

Although electrical energy can be generated in many ways, the most reliable and efficient method has used gravitational energy in the form of falling water. The production of electricity from water is called **hydro-electric energy**. Hydro-electric power is a clean form of energy and an excellent energy source used widely in Ontario and Quebec.

Hydro-electric generating stations take up an enormous amount of land. Often a dam must be built, which causes the land above the dam to be flooded. The dammed water is then released through controlled channels. This racing water turns huge turbines, which turn generators that create electricity. The electricity is carried through huge transmission cables supported on high towers.

It is important to consider some of the drawbacks of hydro-electricity. While this source of energy is clean and renewable, much land is lost in the flooding. Hundreds of hectares of land are often affected, including all human, plant, and animal life living on them.

In recent years, some Native peoples of Canada have launched legal actions protesting against hydro-electric projects. They are protesting against the unalterable changes made to areas that used to be their natural hunting and fishing grounds. Naturalists have also pointed out the long-term negative effects on the environment that result from the flooding of large areas.

How Energy Sources Are Used

Many machines will work only if they have some source of energy. The problem for designers is to find the most inexpensive source of energy that is environmentally safe.

Combustion Engines

The **steam engine** uses burning coal — a fossil fuel — to convert water into high-pressure steam. This was one of the first combustion engines. **Combustion** is the burning of a material using oxygen, or air. Combustion engines harness the heat they produce. In a steam engine, the burning of coal happens in an area outside the actual engine. This is why this type of engine is called an external combustion engine.

The confined steam forces a turbine to rotate or a piston to move. A piston is a thick, sliding disk fitted inside a hollow cylinder. The disk is fitted to a hinged rod attached to a rotating shaft. As the pressure from the steam increases, it forces the piston down, causing the shaft to rotate.

Steam engines were used in the nineteenth and early twentieth centuries. However, they were too heavy and inefficient, and they produced polluting smoke. Eventually they became obsolete, and they are rarely used today.

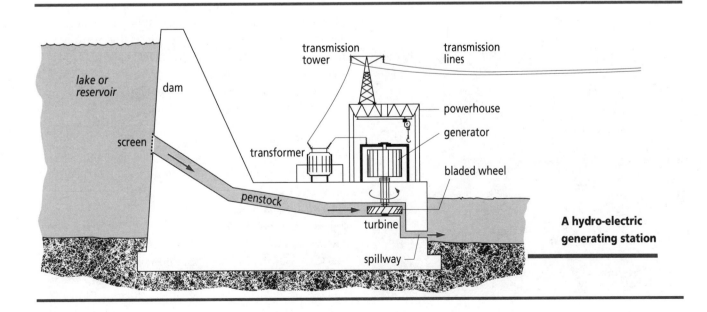

A hydro-electric generating station

**How the
internal
combustion
engine works**

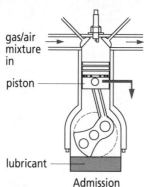
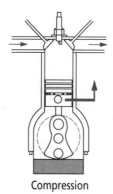
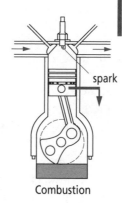
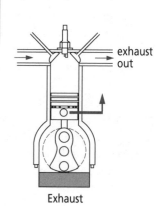

gas/air
mixture
in

piston

spark

exhaust
out

lubricant

Admission   Compression   Combustion   Exhaust

Machines may also be powered by a chemical reaction ignited by a spark, driving a piston to turn a shaft. This is the principle behind the internal combustion engine. In this type of engine, the fuel is burned inside the engine, the parts are smaller, and the fuel lasts longer.

Distillation of oil — another fossil fuel — led to the discovery of gasoline. In an internal combustion engine, a spark ignited in a closed cylinder containing a piston attached to a crank shaft causes the gasoline vapour to explode and drive the piston down. Weights on the crank shaft bring the piston to the top to refire again, causing the shaft to make a complete revolution.

**In the Shell Fuelathon,
engineers put their
new car designs to the
test. These vehicles
use an internal
combustion engine.**

Power Generators

Water that is moving and falling contains gravitational and kinetic energy. Water that is dammed can be forced to fall great heights, causing it to move at high speeds. The fast-moving water carries a great deal of potential energy.

A device called a **generator** harnesses this energy. The falling water strikes blades, causing the generator shaft to turn. A magnet on the centre of this shaft turns as well. The magnet, spinning in a coil of wire, releases particles called **electrons** into the wire surrounding the shaft. Electrons are the small, moving particles that surround the nucleus of an atom. They are picked up and passed from one atom to the other throughout the wire. This creates electrical energy. Electrical energy that flows along a path is called **electricity**.

ELECTRICITY
Electricity can be compared to flowing water. When you turn on a tap, you see the water flow. You can touch the water and feel and hear its force.

Similarly, electricity is a stream of electrical charges that flow along wires. You cannot see this flow, which is called a **current**. However, when a light is burning brightly, you know that the electricity is flowing and is completing a **circuit**, or path, along which the electricity flows.

Materials that allow electricity to flow through them are called **conductors**. For example, metal is a conductor. Materials that allow precise control of electrical current are called **semiconductors**. For example, silicon is a semiconductor.

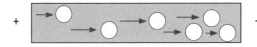

+   –

The benefits of electricity are too numerous to list. Can you imagine what life would be like without it? But electricity can also be very dangerous, whether it comes from a lightning bolt or an electrical outlet. You must be very careful around electricity, for even a small shock can do damage.

We have yet to discover a way to harness the energy in lightning

Electricity that comes from generating stations flows through high-voltage transmission wires and eventually reaches our homes. The flow of electricity in these transmission wires creates an electrical force field, which is why the towers that hold these wires are so tall. By various means of control, the force of electricity is reduced to a volume that can be used in our homes.

You are probably aware of a miniature version of the power generator — the bicycle generator. If the shaft of this small generator is fitted with a bladed wheel and held against a moving tire, the bladed wheel spins, turning the magnet, which generates electricity. The electricity flows from the generator through a wire to a light mounted on the bicycle.

A bicycle generator

Electrical Motors

Although the apparatus is almost the same as in power generators, electrical motors run in a different way. Electrical energy is supplied through a wire coil to a magnet, causing the magnet to turn. The magnet is mounted on a shaft, so the shaft turns. The rotating motion of this shaft can be used to power something else.

Flow of electrons in a conducting material create electricity

The best example of a common electrical motor is the one found in cars. Chemical energy from the acid **solution** in the battery creates electrical energy. This travels through wires to the starter, which is an electrical motor. The activated starter generates a high voltage in the wire coil. This voltage causes a spark from the spark plug, which ignites the gasoline vapour, causing combustion in the engine cylinders and starting the engine. In this way, we can see how chemical energy is converted into electrical energy, which is converted into motion.

Electronics

Electricity is the study of the generation, distribution, and use of electrical energy. **Electronics** is the branch of engineering and technology that deals with the design and manufacture of devices that use electricity, such as

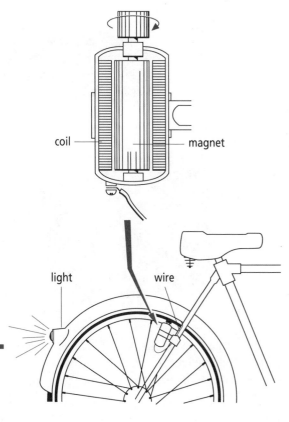

coil   magnet

light   wire

See for Yourself

To see how a small-scale acid battery works, make a lemon battery. Place a copper penny and a zinc nickel in half of a cut lemon (the acid). Attach an electric wire to each coin with metal clips. Using a voltmeter supplied by your teacher, measure the amount of current produced by these dissimilar metals. Make batteries from other fruits and substances. The potential energy of a battery is a method of supplying power to create light, heat, or magnetic energy.

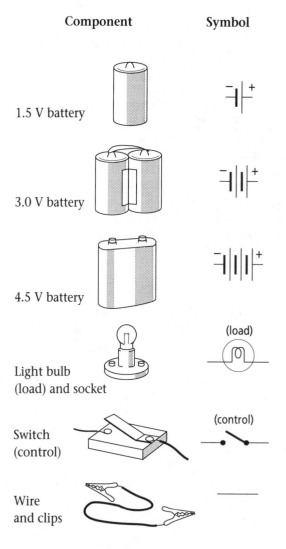

A lemon battery

radios, television sets, and computers. All devices that use electricity have two things in common. The first is that they derive their electricity from a source, such as a battery or a generating station. The second is that these devices form an unbroken circuit through which the electricity flows. A circuit will usually have a **control** (such as a switch) and a **load** (such as a light bulb).

Televisions and VCRs are common examples of electronic devices.

It is important to visualize how electronic components use electrical energy. A simple apparatus and schematic symbols are used to represent these components.

Component	Symbol
1.5 V battery	
3.0 V battery	
4.5 V battery	
Light bulb (load) and socket	(load)
Switch (control)	(control)
Wire and clips	

In a circuit, wires are used to connect a battery to a light bulb. The following diagrams show the apparatus using a common circuit and series and parallel circuits.

| **Apparatus** | **Schematic Diagram** |

1.5 V battery attached to a switch and a light bulb

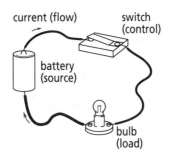
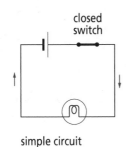

current (flow) switch (control) · closed switch · battery (source) · bulb (load) · simple circuit

1.5 V battery attached to a switch and two light bulbs in a series circuit

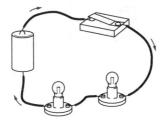
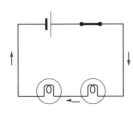

series circuit

1.5 V battery attached to a switch and two light bulbs in a parallel circuit

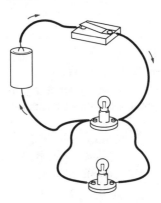
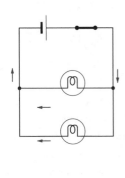

parallel circuit

A **series circuit** is one in which there are two or more loads or controls connected end to end. The same current flows through each of them. There is only one path for the current.

Career Profile

Hugh LeCaine and the Synthesizer

While growing up in Thunder Bay, Ontario, Hugh LeCaine showed an early interest in music and science. He built electronic instruments, sang in church choirs, and studied guitar and piano. Later, he would use these interests to design and fabricate new musical instruments that were easy to play.

In 1931, LeCaine received his Master's degree in science from Queen's University. In 1945, he built the first synthesizer, called the Sackbutt. This was a monophonic instrument; that is, it produced one tone at a time. The Sackbutt synthesizer was like an electronic organ; a keyboard and a foot pedal controlled the sound produced. Joy-sticks controlled the basic waveforms, which are the "shapes" of sound. It is unfortunate that the Sackbutt was never manufactured. Synthesizers were later manufactured in the 1970s, and they became popular musical instruments.

In 1954, LeCaine started Elmus Labs, which was the electronic music research department of the National Research Council. His unique way of thinking about music and creating sound led to the design and fabrication of sixteen different types of musical instruments. During his twenty years at Elmus, LeCaine created touch-sensitive organs. In 1955, he built the Special Purpose Tape Recorder, which could play up to ten tapes simultaneously while varying the speed of each.

While working at the NRC, LeCaine started the second electronic music studio in North America at the University of Toronto. He was also involved in nuclear physics and made contributions to the development of radar. LeCaine was awarded honorary Doctorate degrees from the University of Toronto, McGill University, and Queen's University.

QUESTIONS AND EXERCISES
1. What were some of Hugh LeCaine's early interests?
2. What did Hugh LeCaine do while he was working at the National Research Council?

Conduct an experiment with resistors. You will need a 1.5 V battery, a lamp, a 100 Ω resistor, and three wires. Construct a circuit to light the lamp. Note the brightness. Now connect the 100 Ω resistor to the lamp circuit and note the brightness. Repeat the procedure using different resistors.

A **parallel circuit** is one in which each control or load can operate independently from the other(s). This parallel system is very similar to the electrical lighting in your home, the only difference being that the source of energy comes from outside the home and the circuit is controlled by a switch. When the switch is open, the circuit is broken; when the switch is closed, the circuit is complete.

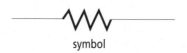

symbol

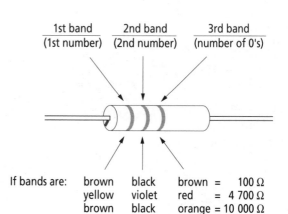

| 1st band (1st number) | 2nd band (2nd number) | 3rd band (number of 0's) |

If bands are:

brown	black	brown	= 100 Ω
yellow	violet	red	= 4 700 Ω
brown	black	orange	= 10 000 Ω

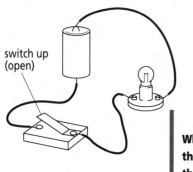

switch up (open)

When the switch is open, the circuit is broken and the bulb does not light up.

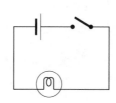

Number	Colour
0	black
1	brown
2	red
3	orange
4	yellow
5	green
6	blue
7	violet
8	grey
9	white

Whether the end result is to make a light bulb glow or to make an alarm bell ring, most electronic devices work in the same way. An electric current is introduced and modified through electronic components in order to operate a device. Aside from the most simple examples of electronic components, there are others that are equally important to the understanding of modern electronics.

Components

RESISTOR

The **resistor** is a component that reduces the current, or flow, of electricity in a circuit. Many electronic components are delicate and therefore can be damaged by large currents. The resistor helps control the flow, thus producing a smaller current. Resistors protect delicate components and control the voltages at different points in a circuit. Most resistors contain carbon, which is a poor conductor of electricity. Resistance — the property that restricts the flow of electricity — is measured in ohms (Ω) and is indicated by coloured bands.

The **potentiometer** is a variable resistor. The most common variable resistor is the one used for the volume controls on audio devices such as radios and stereos.

The **light-dependent resistor (LDR)** relies on the fact that light can increase

Use this diagram of a resistor and the chart to help you calculate resistance given different combinations of coloured bands.

the number of free electrons in a semiconductor. In an LDR, the semiconductor forms a zigzag between two pieces of metal. As soon as light falls anywhere on the semiconductor, it produces free electrons. This allows the current to flow more freely between the two pieces of metal. More light lowers the resistance of the semiconductor. The LDR is used to turn on streetlights in the evening. The streetlights may also turn on on very overcast days due to the lack of light.

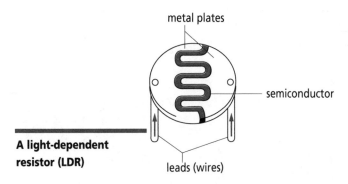

A light-dependent resistor (LDR)

DIODE

A **diode** is a semiconductor that allows current to flow in one direction only. It has a similar appearance to a resistor, but a diode has only one dark-coloured band near one end. The band and the arrow show the conducting direction. To make a diode, special impurities are added in a process called doping. Phosphorus is used to dope the end that has the coloured band. This is the N-type semiconductor. The other end is doped with boron to produce a P-type semiconductor.

A **light-emitting diode** (LED) emits light when a current flows through it. LEDs are often used in a circuit to show that current is flowing. They may emit infra-red light or visible light, and they can be used as indicators in place of light bulbs. LEDs last longer than light bulbs and use much less power. The glowing lines you see on digital clocks are LEDs. LEDs are also used to light up instrument panels, such as on the dashboard of a car.

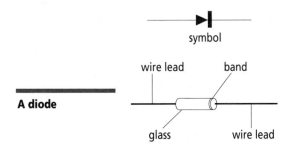

A diode

CAPACITOR

A **capacitor** stores electricity like a battery. There are two kinds of capacitors: the ceramic disk capacitor stores a small charge; the electrolytic capacitor stores a large charge. A camera flash gun uses a capacitor.

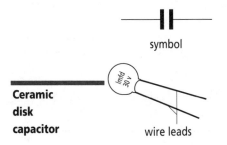

Ceramic disk capacitor

 Caution: *You can get a nasty shock from electrolytic capacitors in high-voltage equipment. This can happen even when the equipment is no longer plugged into the power supply.*

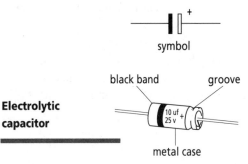

Electrolytic capacitor

TRANSISTOR

A **transistor** acts as a fast switch, switching current on or off, or amplifying (increasing) a small current. The invention of transistors in 1952 made it possible to build battery-powered portable radios.

The transistor must be correctly connected or it will be damaged. Each lead must be identified as an emitter, a collector, or a base. The *emitter* (a) allows electrons into the transistor. It is usually connected to the negative side of both input and output circuits. The *collector* (c) is farthest from the emitter terminal. It collects electrons from the transistor. The collector is connected to the positive side. The *base* (b) is the central section. The input current is connected to this terminal. It can take only a very small current and therefore needs protective resistance in the form of a resistor.

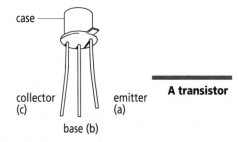

A transistor

A transistor can be made to switch on and off by applying a small current to a base leg. When there is no current flowing to the base, there will be no current flow between the collector and the emitter. A small current to the base allows a large current to flow between the emitter and collector. The large current can be used to power electronic components.

There are two types of transistors: NPN and PNP. The most common type is NPN. The plan for your electronic project will indicate the type of transistor required.

Nowadays we don't use transistors as much as we do integrated circuits.

INTEGRATED CIRCUITS (ICs)

An **integrated circuit** (IC), or microchip, contains thousands of miniature electronic components in a thin piece of silicon seldom larger than a match head. Yet, ICs can be manufactured as cheaply as transistors, and they use very little electricity. They are usually mounted in larger black boxes with two rows of terminals to make them easier to handle. Today, manufacturers of integrated circuits are competing with one another to produce more and more complex and useful circuits on smaller and more affordable silicon chips. These components are the basis of computers.

The Computer

Probably the most common sophisticated use of electronics is in the design and manufacture of computers. The uses of computers are incredibly varied. This complex array of electronic circuits allows for quick storage and change of data in an electronic format.

The personal computer has changed the way the average person does his or her daily work. It enables people to do word processing, mathematical calculations, and drawings quickly and easily. In fact, this book was designed and produced on a computer. There is no doubt that the use of a computer can increase your productivity in most areas of work.

Like all other technology, however, steady use of a computer may cause health problems. Although much research has been done on these health risks and no definite results have been found, there is still concern over the magnetic field that is emitted by the computer. Also, the rays that are emitted from the computer monitor, like those from a television, may be a health risk.

Energy and Technology

So far in this chapter you have learned how energy is important in design and technology. For example, in the section on electrical motors (page 93) you discovered a common way in which energy is converted, using technology, into a useful product. Eventually, products themselves become technology, since they can be used or modified to fulfill other needs. For example, the use of design, technology, and energy helped to create the car at the beginning of the twentieth century, fulfilling the need for transportation. Eventually, the car was thought of as another type of technology that fulfilled other needs.

The following table shows energy use and the development of major technological discoveries from the dawn of time to the present.

Energy Producer	Uses and Benefits
Sun	• warmed the Earth • plants grew, animals ate plants, animals were eaten by humans
Fire	• could be contained within a circle of rocks • provided light, warmth, and heat for cooking and defence
Animal	• controlled by a rope, harness, or yoke • carried riders and pulled loads • not an energy producer, but a major technological discovery that led to the development of other technology
Wheel	• moved large loads • transported more people and loads
Sail	• controlled by its angle to the wind • reduced human energy • attached in groups on a wheel to make a windmill • allowed new lands to be discovered
Water	• waterwheels turned gears and machines
Steam	• powered labour-saving machines • replaced the sail, animal power, and wind power • marked the industrial age • introduced new energy sources
Coal	• burned longer and hotter than wood • used in factories, on ships, and in wheeled vehicles

Alternative energy

Nuclear
- splitting of atom (nuclear fission)
- steam created by the water that cools the reactor is used to turn the turbine that generates electricity

Solar
- development of photovoltaic cells
- converts electromagnetic radiation into electrical energy

Wind
- wind-driven generators produce electrical energy

Hydro-electric power
- led to development of the generator
- controlled by confinement in insulated copper wire
- replaced steam, gasoline, and water power in factories
- efficient source of electricity for homes and industry

Internal combustion
- fuel burned inside the engine
- engine parts could be smaller and engine lighter
- replaced all steam applications on land and sea
- made flight possible

Petroleum
- discovery of crude oil
- distillation process creates kerosene for fuel
- provides gasoline for the internal combustion engine

Points for Review

- Energy is the capacity for doing work. The common forms of energy are kinetic (active) and potential (stored).
- Gravitational energy is created by the motion of a falling object.
- The following types of energy are created from secondary sources: thermal, radiant, sound, electrical, magnetic, chemical, and nuclear energy.
- The availability of a source of energy determines its cost.
- Fossil fuels are non-renewable sources of energy. The burning of fossil fuels is a major contributor to the greenhouse effect and global warming.
- Solar energy is a renewable resource that can be collected, stored, and converted using solar collectors and photovoltaic cells.
- The wind is a renewable source of energy that is used to turn the blades of windmills and turbines to generate electricity.
- A nuclear reactor uses uranium to create energy through the process of fission.
- The production of energy from falling water is called hydro-electricity.
- The internal combustion engine uses the heat from burning fuel within the engine to run.
- Electricity is the study of the generation, distribution, and use of electrical energy.
- Electronics deals with the design and manufacture of devices that use electricity.
- All electronic devices derive their electricity from a source and form a circuit through which the electricity flows.
- Electronic circuits usually have a control and a load.
- The computer has changed the way the average person does his or her daily work.

Terms to Remember

alternating current (AC)	electronics	light-emitting diode	radiation
capacitor	electrons	(LED)	radioactive
chemical energy	energy	load	renewable
circuit	fossil fuels	magnetic energy	resistor
combustion	fractional distillation	natural gas	semiconductors
conduction	generator	non-renewable	series circuit
conductors	gravitational energy	nuclear energy	solution
control	greenhouse effect	nuclear fission	sound energy
current	hydro-electric energy	parallel circuit	steam engine
diode	integrated circuit (IC)	photovoltaic cells	thermal energy
direct current (DC)	kinetic energy	potential energy	transistor
electrical energy	light-dependent resistor	potentiometer	uranium
electricity	(LDR)	radiant energy	

Applying Your Knowledge

1. Draw one illustration showing all the sources of energy that are found in nature. Be sure to write beside each energy source what it is.
2. Explain in a brief paragraph what the term non-renewable resource means. Which of the following could be considered non-renewable: soiled laundry, trees in the forest, coal?
3. List the pros and cons of nuclear and solar energy. In a small group, discuss your list.
4. Using a decision-making chart (see Chapter 2, page 19), list five different ways of creating energy. Write the environmental factors that are affected in the spaces across the top of the chart. Now rate the energy sources to find out which is the most environmentally friendly.
5. What form of energy is the driving force for the car? In a group, discuss and list the benefits and drawbacks of cars. Discuss other ways they might be powered.
6. Explain briefly what is meant by the greenhouse effect. (HINT: Your work in question 5 may help.)
7. What climate and environmental conditions would create the need to find an alternative energy source to hydro-electric power? Make a list of these conditions, and then rank them from most likely (number 1) to least likely.
8. To discover the importance of electricity, try to think of any room in your home that does not have at least one electrical appliance. Make a list of your favourite extra-curricular activities. Place a check mark beside the ones that depend on electrical energy.

9. Make a list of the ways you can reduce energy consumption.
10. List the pros and cons of the following with respect to energy consumption and effect on the environment:
    a. walking,
    b. riding a bike,
    c. driving a car.
11. Think of at least three machines that use LEDs. List these machines and describe how the LEDs are used.
12. Sketch the following:
    a. a 1.5 V battery connected to a light bulb, with the switch on,
    b. a 3.0 V battery connected to two light bulbs, with two switches (one open and one closed),
    c. a 1.5 V battery connected to two light bulbs, with one switch closed.
13. Without rereading the relevant part of the chapter, tell which of the following is the most common transistor:
    a. PNP,
    b. PCB,
    c. NPN,
    d. LED.
14. Complete the steps in the design process to fulfill the following need. Make sure to write down the requirements of each step.
    Need: To design something to amplify sound.

Chapter 7

MACHINES

What You Will Discover

After completing this chapter, you should be able to:

- Understand the importance of machines in everyday life.
- Identify simple machines.
- Design and fabricate something that applies mechanical motion.
- Understand the basics of power transmission.
- Identify how simple machines are used to make a robot.

*I*ndustrially and technologically advanced nations of the world are founded on **machines**. Machines are devices made up of fixed and moving parts that do specific jobs. They are an integral part of the world we live in. The development of machines has enhanced what we can achieve with our hands and unaided physical strength. Look around you, and you will see the range of machines that have become part of our way of life.

There are many machines for doing household chores. The kitchen contains machines from the simple can opener to the microwave oven. We mix cake batter with a spoon (a simple machine) or with a food processor (a complex machine).

Transportation also involves machines. The bicycle, the car, the snowmobile, the boat, the train, and the airplane are all machines. It is difficult to imagine a world without these means of transportation.

An oxcart

A wagon train

An early automobile

A train locomotive

A modern automobile

Robots are machines used by police forces to assist them in dangerous situations. Other robots are used in factories to perform repetitive or hazardous work.

Prostheses are machines used by people who require hand, arm, or leg replacements. These artificial limbs cannot completely replace human limbs, but they help their users to lead more active lives. Powered wheelchairs are other machines that enable the physically challenged to move about on their own.

Prostheses are machines that help people who have missing limbs.

Space probes are machines that are sent to other planets to carry out research and testing. They have made it possible for humans to obtain valuable information from outer space without having to risk their own safety.

The lunar rover was used for transportation on the moon.

All of these types of machines have been designed and fabricated using technology to fulfill needs. They enable work to be done in more efficient ways. More and more, people also consider leisure not just a "want" but also a need. Our minds and bodies require the physical and mental stimulation of leisure. Many machines have been designed to fulfill these leisure needs.

A baseball bat is a machine that provides an advantage when hitting a ball. If you were to try hitting a ball with your arm, you would realize the value of using a baseball bat. A hockey stick is also a machine, as it is used to propel a puck. A tennis racket is another machine.

A hockey stick is a simple machine; a Ferris wheel is made of several machines.

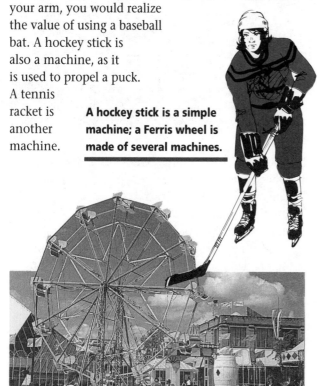

People and Machines

The human ability to make use of the forces and materials available in nature is nowhere more apparent than in the range of machines at our disposal. Our lives have been transformed by the innovative skills of the designer and the fabricator in the development of machines.

Many machines operate efficiently at the push of a button, but there are many other machines that require their operators to be skillful and mindful of safety.

Often the skill required to develop and operate a machine is not fully appreciated. The successful design and installation of a large industrial machine, for example, requires the sustained effort of a team of experts. Procedures for assembly, inspection, and testing must be carefully executed and completed before the machine can be declared to be in safe working order.

It is important to realize that the design and construction of a machine rarely makes people unnecessary to a job. Although machines may perform simple tasks, unskilled people are replaced with skilled machine operators. Other people may be needed to repair the machines and keep them running smoothly. Unskilled people may need to be retrained to do the new jobs. In this way, machines eliminate old jobs and create new ones.

The Canadarm is an example of a machine that took many years of planning and testing by several teams of workers before its debut in space. Designed as part of the space shuttle, it has to withstand the liftoff and perform its function as a robotic arm flawlessly in space.

The Canadarm took years of teamwork to develop.

Why You Need to Know about Machines

As a designer, you need to know in a general way what machines can do. You need to understand how machines have been used in the past and present, and how they might change the future. Machines fulfill the following general needs:

- they provide a mechanical advantage when performing a task,
- they replace human or animal effort at performing physical tasks,
- they increase forces and change the direction of motion,
- they maintain accuracy in performing a task,
- they lower the cost of performing a task,
- they reduce the drudgery of performing a repetitive task,
- they reduce the time taken to perform a task,
- they can work in dangerous environments,
- they eliminate old jobs but create new ones.

As a design student, you will need to know about mechanical advantage, mechanical movement, and machine application. **Mechanical advantage** is the ability to move a large load using a small effort, or force (see Chapter 5 for more on load and force). This knowledge will enable you to decide which machines to use to move, drive, or power a project. The remainder of this chapter will focus on mechanical movement and machine application.

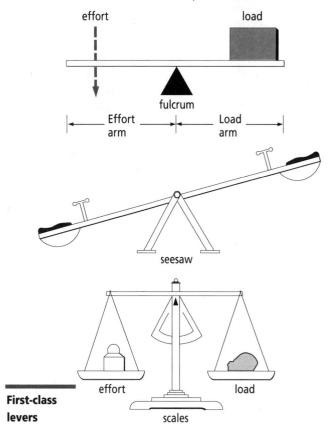

First-class levers

Simple Machines

*T*he main reason for using a machine is the mechanical advantage it gives us. Simple machines provide the means of supplying the mechanical advantage. There are six simple machines: the lever, the wheel and axle, the pulley, the inclined plane, the wedge, and the screw.

The Lever

The **lever** is a bar that magnifies the effect of a small **effort** (the applied force) to move a large mass called the **load**. The bar turns on a fixed point called the **fulcrum** (or pivot). There are three basic types of levers: first-class, second-class, and third-class.

In a **first-class lever**, the "fulcrum" is placed between the *effort* and the *load*. Examples of the first-class lever are the oar, the seesaw, and the scale. The first-class lever can be doubled to produce other useful machines, such as scissors, pliers, and vise grips.

In a **second-class lever**, the "load" is between the *effort* and the *fulcrum*. Examples of the second-class lever are the wheelbarrow, the crane, and the pry bar. The second-class lever can be doubled to produce machines such as the nutcracker.

Second-class levers

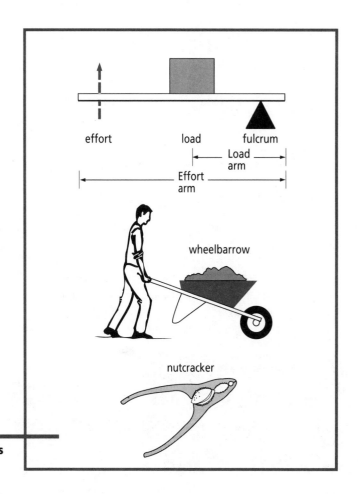

In a **third-class lever**, the "effort" is between the *load* and the *fulcrum*. Examples of the third-class lever are the shovel, the field-hockey stick, and the baseball bat. Some of your favourite sports use these levers to gain a mechanical advantage. The third-class lever can be doubled to produce machines such as tweezers.

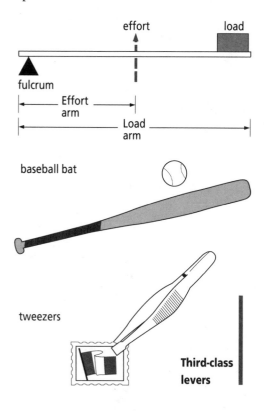

effort   load

fulcrum

Effort arm

Load arm

baseball bat

tweezers

Third-class levers

Effort or force is amplified — made greater — when the load is placed closer to the fulcrum. A small effort applied far from the fulcrum exerts the same amount of moment as a large effort applied closer to the fulcrum. **Moment** is the product of the effort multiplied by the distance from the fulcrum.

The Wheel and Axle

The **wheel and axle** is a machine that we use every day. Buses, trucks, cars, bicycles, and roller skates all have wheels that rotate around an axle.

The wheel and axle has developed progressively in different types of machines. The success of the wheel and axle depends on a rolling contact between parts to keep on turning indefinitely. Although the basic structure of a wheel and axle remains the same, there are two types of this simple machine:

1. A wheel and axle transmits force to do some work. Effort applied to the wheel causes the axle to turn with greater force than the wheel. An example is the rear wheel of your bicycle when you transmit force by pedalling.
2. A friction-fighting wheel does not transmit force, but only reduces friction against another surface. An example is the front wheel of your bicycle, which rotates over the ground, thus reducing friction.

Most commonly, you will recognize the first type of wheel and axle. For example, a bicycle has a number of wheels and axles. The chain that connects the driving sprocket (attached to the pedals) to the driven sprocket (on the back wheel) is a specialized wheel and axle.

How many wheel and axle machines are there on this toy biplane?

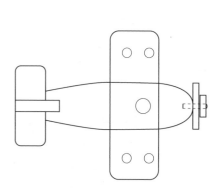

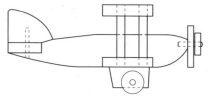

See for Yourself

How many wheel and axle machines can you find on a bicycle or tricycle? Count the total number of wheel and axle combinations on one of these cycles. Make a list of where they occur, and compare your findings with a classmate.

FRICTION

The resistance to motion between two surfaces that touch is called **friction**. Wheels help overcome friction, since only a small part of a wheel comes in contact with a surface at any time. It is important to understand friction because many types of complex machines use this force in order to work.

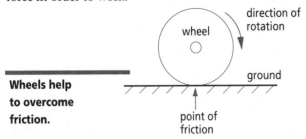

Wheels help to overcome friction.

Friction allows a bicycle to move forward. The friction between the ground and the tires allows the force exerted by the pedals to be transmitted to the ground. Too much friction impedes movement (slows you down). On the other hand, movement is nearly impossible if there is too little friction, as when you try to ride on ice. An example of useful friction is the brake on a bicycle. The brake increases friction when it grabs the wheel; the wheel, in turn, slows down.

Friction can be reduced by using lubricants and bearings. You lubricate the axles of your bicycle with oil or grease to keep them moving freely. If you take apart the hub of a bicycle wheel, you will see the ball bearings that surround the axle. When properly lubricated, these bearings make it easier for you to pedal your bicycle because there is less friction.

The Pulley

The **pulley** is a wheel that transmits force by means of a belt, rope, or chain passing over its rim. Pulleys are used on fishing boats to haul in the nets, in the logging industry to move cut timber, and in the construction trades to move materials from one level of a building to another. Elevators in highrise buildings and in mines use pulleys to gain a mechanical advantage.

The mechanical advantage of a single fixed pulley can be calculated as follows:

$$\text{Mechanical Advantage} = \frac{\text{Load}}{\text{Effort}}$$

A single pulley in the down position is said to have a mechanical advantage of 1. That is, the force required to move the load is equal to the mass of the load. A single pulley in the suspended position is said to have a mechanical advantage of 2. This means the pulley uses one-half the force to lift the load. The load is equally distributed between the support and the effort.

Pulleys are used to pull in fishing nets.

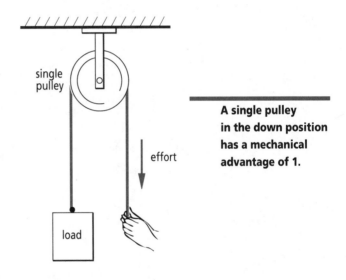

A single pulley in the down position has a mechanical advantage of 1.

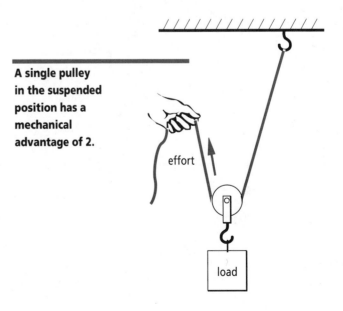

A single pulley in the suspended position has a mechanical advantage of 2.

The mechanical advantage of a pulley system increases with a more complex arrangement. One method of determining the mechanical advantage of a complex pulley system is to count the sections of rope supporting the load.

Suppose a pulley system with a mechanical advantage of 1 is used to lift a load a distance of 1 m. One metre of rope will have to be pulled in to lift the load 1 m. When lifting the same load to a height of 1 m using a pulley system with a mechanical advantage of 4, the operator must pull in 4 m of rope. However, with this system the operator uses only one-quarter the amount of effort used in the first system.

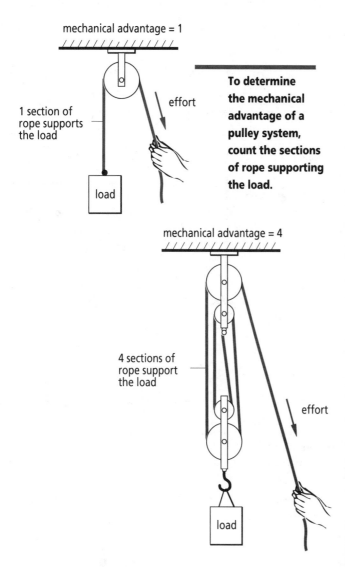

mechanical advantage = 1

1 section of rope supports the load

effort

load

To determine the mechanical advantage of a pulley system, count the sections of rope supporting the load.

mechanical advantage = 4

4 sections of rope support the load

effort

load

The Inclined Plane

The **inclined plane** is a machine that provides a sloping surface to gain access to lower and higher levels. It is a fixed machine — it does not move. The inclined plane

Career Profile

Olivia Poole and the Jolly Jumper

Many parents today are thankful for the Jolly Jumper. They know that this device helps promote the physical development of an infant while entertaining him or her. What parents may not know is that the Jolly Jumper was developed by a Canadian: Olivia Poole of British Columbia.

As a mother, Poole wanted to find a way for an infant to get exercise without being held by a parent. Poole's Ojibway heritage had introduced her to the cradle board, which hung from the flexible branch of a tree. The infant was strapped in, allowing his or her feet to just touch the ground. With a kicking motion, the infant could jump up and down. The apparatus helped develop leg muscles and release pent-up energy.

Poole designed her first Jolly Jumper using materials in her home — a broom handle, a pillow, and a coiled spring. In 1953 she rented out Jolly Jumpers for one dollar a month. By 1959, she was overseeing the production of several thousand Jolly Jumpers a month in North Vancouver. The Jolly Jumper now has a more sophisticated design using linkages and pulleys. This product is a household name not just in Canada, but also in the United States, Australia, and Britain — thanks to an innovative design developed by Olivia Poole.

QUESTIONS AND EXERCISES

1. Describe how Olivia Poole's Ojibway heritage inspired her to design the Jolly Jumper.
2. What materials did Olivia Poole use to create her early Jolly Jumper design?
3. Sketch five versions of another product that infants could use.

does not alter the amount of work that is needed, but it alters the way in which the work is done. A bridge over railway tracks is an inclined plane. Inclined planes make it easier for cars and trucks to ascend or descend hills.

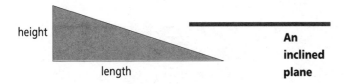

An inclined plane

Inclined planes are now a common feature as ramps in buildings and on sidewalks. They make it easier for people in wheelchairs to move around. The sloping surface provides a slow descent as opposed to a sudden drop over a steep edge. How fast an object moves down a ramp is controlled by friction and by gravity, the force that keeps objects fixed to the surface of the Earth.

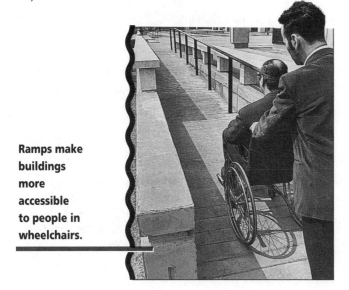

Ramps make buildings more accessible to people in wheelchairs.

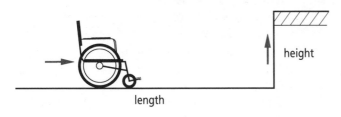

How a ramp works

The Wedge

The **wedge** is an inclined plane with one minor variation: it moves. The cutting edge of an axe or a chisel is a wedge. The cutting teeth on a band saw or table saw blade are wedges. Materials are cut by the wedging action of the saw teeth. More complicated devices, such as the zipper and the can opener, are based on the wedge.

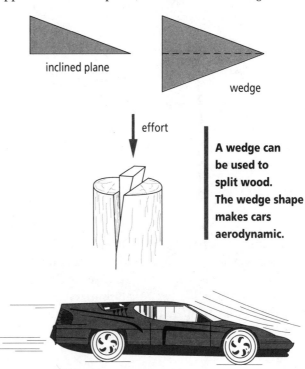

inclined plane   wedge

A wedge can be used to split wood. The wedge shape makes cars aerodynamic.

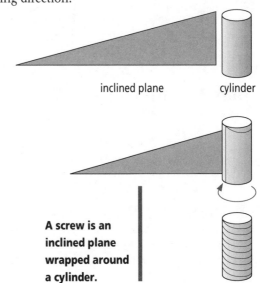

The Screw

The **screw** is an inclined plane wrapped around a cylinder. As a screw is turned, the spiral thread moves away from or toward the turning effort, depending on the turning direction.

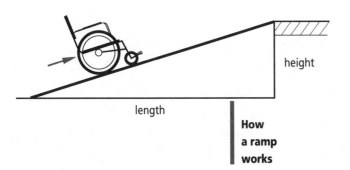

inclined plane   cylinder

A screw is an inclined plane wrapped around a cylinder.

The distance between the screw threads is called the **pitch** and equals the distance the screw travels when it is turned once.

There are many different kinds of screws. A bolt is a screw. A bench vise contains a screw. A car jack also uses a screw to lift a car.

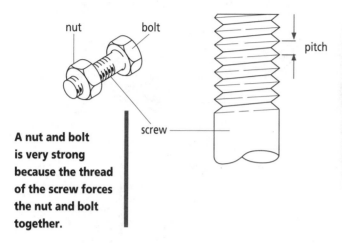

A nut and bolt is very strong because the thread of the screw forces the nut and bolt together.

An **Archimedes' screw** is a broad, threaded screw encased in a cylinder that is used to raise water from one level to another. It is named after its Greek inventor. An **auger** is another, similar screw that is used on farms for transporting feed to different levels or along a horizontal plane. An auger bit is used to bore holes through wood.

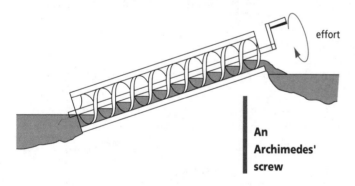

An Archimedes' screw

Complex Machines

*B*ased on the concepts of simple machines, more complex machines have evolved to enable force to be transmitted with greater efficiency and flexibility.

Uses of Wheels and Axles

It is impossible to list all the uses of the wheel and axle in complex machines, but a few are described here as examples.

A very sophisticated use of the wheel and axle can be found in the jet engine. Blades have been added to the wheel, which is connected to an axle that rotates to produce tremendous power for high speed and liftoff. The wheel and axle is used on the hovercraft to lift it off the water and move it forward.

You can use the wheel and axle in a variety of interesting projects. Toys can be designed using four, three, two, or even one wheel. You can construct axles from metal rods or wooden dowels.

The propellers on a hovercraft are complex forms of the wheel and axle.

Power Transmission

Wheels can be used with other wheels to transmit force from one axle to another axle, or from one shaft to another shaft. To **transmit** means to pass along energy. This movement distributes the mass of a load.

There are four methods of power transmission that use wheels. They are: the wheel to wheel, the belt drive, the chain drive, and the gear drive.

There are three principles of power transmission:
1. a small driver wheel pushing a large driven wheel will give low speed and high force (**torque**),
2. a large driver wheel pushing a small driven wheel will give high speed and low force, and
3. a driver wheel and a driven wheel of the same size will have the same speed and force.

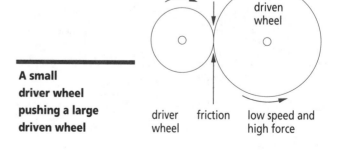

A small driver wheel pushing a large driven wheel

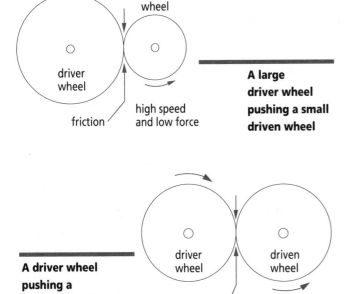

driven wheel

driver wheel

friction

high speed and low force

A large driver wheel pushing a small driven wheel

A driver wheel pushing a driven wheel of the same size

driver wheel

driven wheel

friction

same speed and same force

THE BELT DRIVE

The **belt drive** is the simplest means of transferring power from one wheel to another wheel. These wheels are pulleys — that is, they are usually grooved so they can be fitted with a belt. The goal is not just to transmit motion from one pulley to another, but also to their connecting axles. These axles are called **shafts**.

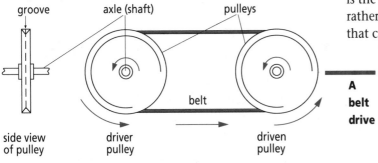

groove

axle (shaft)

pulleys

belt

side view of pulley

driver pulley

driven pulley

A belt drive

The belt transmits the power from the driver pulley to the driven pulley. By changing the size of the driven pulley, we can change the speed at which the shaft rotates. A large pulley driving a small pulley gives high speed and low force. Conversely, a small pulley driving a large pulley gives low speed and high force.

By crossing the belt, the direction of the driven pulley can be reversed. The pulley and belt arrangement can be further refined for different applications. You will find many advantages in using the belt and pulley in your designs. They are easy to install and inexpensive to purchase.

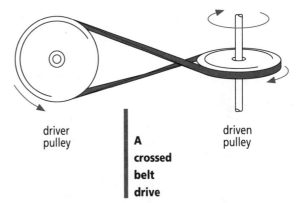

driver pulley

A crossed belt drive

driven pulley

THE CHAIN DRIVE

Bicycles move by means of a **chain drive**. The driver sprocket is attached to the crank, and the driven sprocket is the back wheel. The use of toothed wheels and a chain rather than a belt eliminates the possibility of slipping that can occur in a belt drive.

A chain drive is used in the caterpillar tracks of a bulldozer or an excavator. Such tracked vehicles are capable of negotiating slopes of 20° to 25°; conventional wheel-mounted machines are limited to 15° to 20° slopes. They also provide a greater pulling force for moving bigger loads.

See for Yourself

To understand how driver and driven wheels operate, construct from cardboard and dowels a 5 cm wheel on an axle and a 3 cm wheel on an axle. Both wheels should be able to rotate on their axles. Using a glue gun, mount the axles on a piece of cardboard so that the wheels touch. Try turning the larger wheel four times, and then six times. How many revolutions did the smaller wheel turn? Now try turning the smaller wheel the same number of times. How many revolutions did the larger wheel turn? Note how you think driver and driven wheels work, and present your observations in a class discussion.

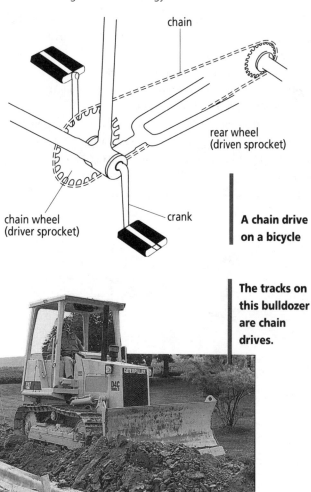

chain

rear wheel
(driven sprocket)

chain wheel
(driver sprocket)

crank

**A chain drive
on a bicycle**

**The tracks on
this bulldozer
are chain
drives.**

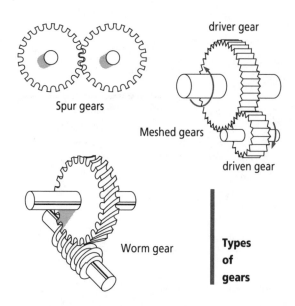

driver gear

Spur gears

Meshed gears

driven gear

Worm gear

**Types
of
gears**

Chain drives have many advantages. They are more efficient because they do not slip; they are flexible when properly lubricated; they can withstand temperature changes and dirt; and they can support greater loads than belt drives.

The principles of the chain drive are basically the same as those of the belt drive. You can install a chain drive on a project just as easily as a belt drive. However, the chain drive costs more and is harder to find, although most hobby shops carry chain drive components.

THE GEAR DRIVE

A gear drive is an assembly of gears. **Gears** are wheels with teeth. They are used to transmit force from one part of a machine to another. Gears are a great improvement over a conventional driver wheel, which uses only friction to transfer the force to the driven wheel.

Each tooth of the driver gear acts as a lever that transfers force to each tooth of the driven gear. This combination of machines makes the gear capable of transmitting a large amount of power.

A **worm gear** is a type of screw. It is commonly used to transfer power from a motor to a wheel. The worm gear also allows the high turning speed of motors to be contained within a small operating space.

Gears are the most durable and rugged of all mechanical drives. They can transmit force up to 98% efficiency and are found on truck transmissions and on most heavy-duty machine drives. Gears are more expensive than most other power drives, especially if they are made of machined metal rather than plastic. Plastic gears are suitable for toys such as remote-control cars, which are driven by an electric battery motor using plastic gears to transfer the force to the wheels.

In addition to transferring power from one shaft to another, gears can change speed and direction from a horizontal to a vertical movement. In rack and pinion gears, the wheel (pinion) meshes with a toothed rack to allow back and forth motion.

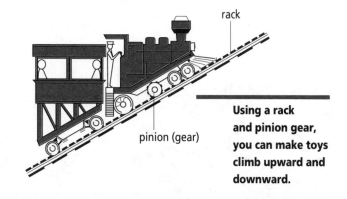

rack

pinion (gear)

**Using a rack
and pinion gear,
you can make toys
climb upward and
downward.**

Gears have many uses. Robots use gears driven by motors to perform tasks such as placing doors on cars in an automotive assembly line. In outer space, the

Canadarm operates with gears. Gears attached to the power source allow the arm to virtually duplicate the movement of a human arm. Prosthetic hands are composed of miniature gears that duplicate human movement.

RECIPROCATING DRIVES
Reciprocating drives change a **rotary motion** (a motion that goes around) to a **linear motion** (a motion that goes back and forth). There are four basic designs of reciprocating drives — the crankwheel and lever, the crankarm and lever, the crankshaft, and the cam.

The **crankwheel and lever** is a wheel with a connecting rod attached to it off-centre. The rotation of the offset connecting rod is transferred into a linear motion. The crankwheel and lever is used to drive the pistons in a motorcycle engine.

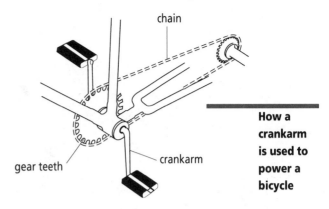

How a crankarm is used to power a bicycle

within the engine casing, so you cannot see it. The crankshaft of a car is simply longer and a little more complex than the crankarm on a bicycle.

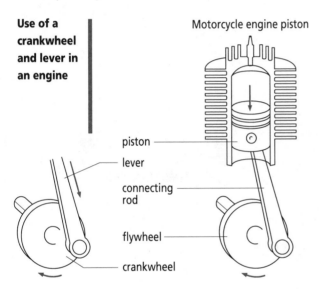

Use of a crankwheel and lever in an engine

Motorcycle engine piston

piston
lever
connecting rod
flywheel
crankwheel

You can easily duplicate the action of the crankwheel and lever by making a model from cardboard. If the model works, you can use a more substantial material such as plastic in your fabricated project. The crankwheel and lever is often used in toys to make them bounce up and down.

The **crankarm and lever** is fundamental to powering a bicycle and the internal combustion engine. A crankarm and lever can turn rotary motion into linear motion.

A crankarm and lever can be used to create movement in toys that you design. A **crank** can be made by bending a piece of heavy-gauge wire. You can make more than one crank on a wire shaft by making more bends in the wire.

The internal combustion engine found in cars, trucks, boats, and trains uses a **crankshaft** to transfer up-and-down motion into rotary motion. The action takes place

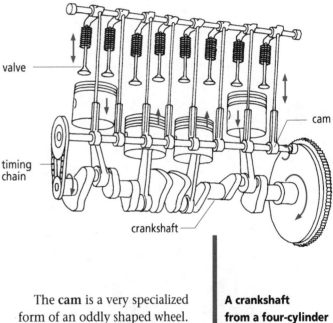

valve
timing chain
crankshaft
cam

A crankshaft from a four-cylinder car engine

The **cam** is a very specialized form of an oddly shaped wheel. It is usually heart-shaped, but it can take a variety of shapes depending on the application. Cams mounted on a shaft rotate, giving motion to a **follower rod**. Cams are found in extremely complex machines, such as the internal combustion engine. They may also be found in less complex machines, such as toys.

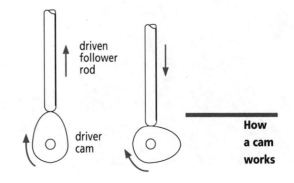

driven follower rod
driver cam

How a cam works

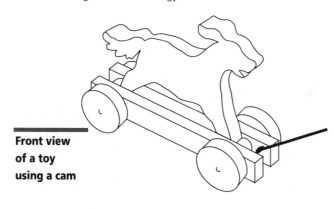

**Front view
of a toy
using a cam**

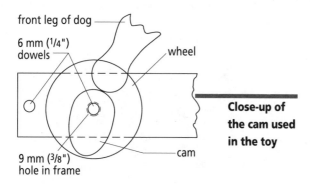

front leg of dog

6 mm (¼")
dowels

wheel

9 mm (³/8")
hole in frame

cam

**Close-up of
the cam used
in the toy**

Linkages

The human body is by far the most complex machine that is known to the human race. We have only to watch how people move to appreciate this fact — the movement of gymnasts, figure-skaters, and hockey players, for instance. Science, through the study of physics, has tried to duplicate the feats performed by the human body.

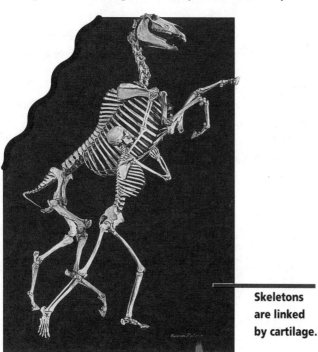

**Skeletons
are linked
by cartilage.**

Designers have used human motion as a model for how a need can be fulfilled. As a result, we have been successful in producing specialized machines, but we are still not able to create a robot that functions exactly like a human.

The human skeleton is an elaborate assembly of linkages of bones, muscles, and joints. An incredible amount of motion is possible between the connected parts. Designers have attempted to duplicate these types of connections in the use of linkages. **Linkages** are parts connected to one another by joints, and they are very important parts of machines. Industry uses the mechanical linkages on assembly lines for production work.

**The legs of
this puppet
are linked
to the body.**

**The linked
arms of this
toy penguin
move.**

dotted
line
indicates
string

puller

BAR LINKAGES

Bar linkages in the steering mechanism of a car enable the transfer of motion to the wheels. The driver selects the degree of movement by turning the steering wheel.

Linkages are used extensively in manufacturing. The aeronautics industry relies heavily on bar linkages to move the wing flaps and operate the landing gear of aircraft. Bar linkages are also used in car-mounted wheelchair lifts.

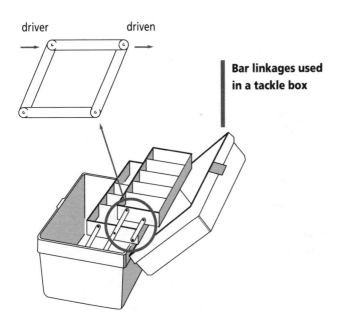

Bar linkages used in a tackle box

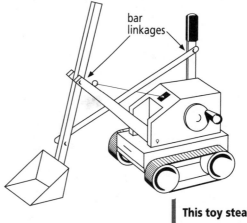

This toy steam shovel uses bar linkages, wheels and axles, and levers.

Bar linkages can be used to make grippers to extend a person's reach. In the past, clerks in stores used grippers to reach the top shelves. Today, physically challenged people use grippers to magnify limited muscle movement and strength.

Bar linkages may be joined together in an X shape. Mobile platforms in the construction industry can be extended using the X-shaped linkages. These make the platforms stable and safe.

A **pantograph** is a specialized X-shaped bar linkage used to reduce or enlarge drawings. For example, in coin design, the coins are first drawn at a large size that shows

Bar linkages are used in heavy construction and excavating equipment. Earth movers and excavators use bar linkages as levers to dig into the ground and move soil or load trucks. The levers are activated by the flow of fluid in hydraulic cylinders, which are in turn operated by a worker. Notice in the photo how the action of the machine's shovel digging into the soil resembles the movement of a human arm and hand.

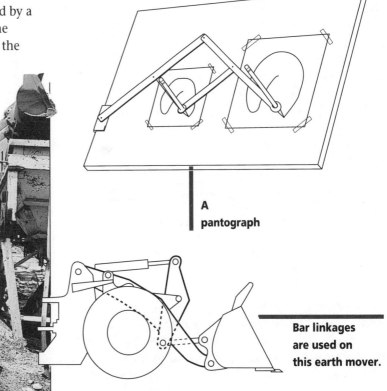

A pantograph

Bar linkages are used on this earth mover.

the fine detail. The pattern is then reduced using a pantograph, which preserves the fine detail on the coin. Similarly, the detailed designs for miniature toy cars are first drawn at a large size and then reduced with a pantograph.

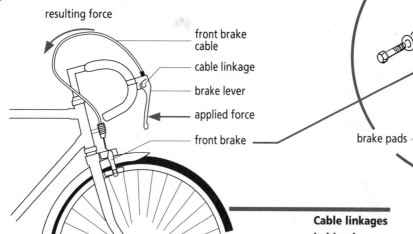

Cable linkages in bicycle hand brakes

CABLE LINKAGES

One of the easiest places to find **cable linkages** is on your bicycle. The brakes may be controlled by the brake lever on the handlebar and a cable that attaches to the **brake calliper**. When you squeeze the brake lever, the cable transfers the force to the brake calliper. The brake calliper presses the rubber brake pad against the rim of the wheel, causing the bicycle to stop.

THE UNIVERSAL JOINT

The **universal joint** is a coupling (a device for joining) that transmits power from one shaft to another shaft that is running in a different direction. This type of joint is usually found on cars and trucks. It is located on the drive shaft, which transfers rotary motion to the rear wheel. This joint also allows for movement up and down and sideways.

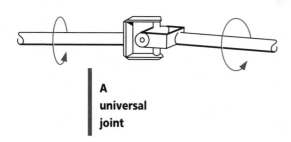

A universal joint

The universal joint was used in the construction of a mechanical wrist that enabled two young Canadians, who were both born with only one hand, to play hockey. The design and fabrication of the prosthesis was completed between 1982 and 1984. It was achieved through the combined efforts and expertise of a mechanical

supervisor, a machinist, and a mechanic. The arm and hockey stick attachment that they created allowed the boys to make wrist shots. The mechanical wrist has given other physically challenged people more agility in their movements.

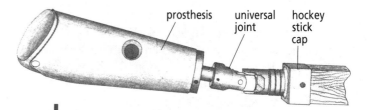

This special prosthesis was designed to allow a person missing one hand to take a wrist shot when playing hockey and can be adapted to other sports. The hockey stick attachment shown here can be replaced with a golf club or a baseball bat. The coupling is a specially modified universal joint.

Source: Designed and drawn by Tom Hennessy, Brunswick Mining and Smelting Corp. Ltd., Bathurst, New Brunswick. Reproduced by permission.

Perhaps the most advanced use of the universal joint can be found in the Canadarm, which uses a series of complex gear boxes to make up a high-precision universal joint.

The ball joint, a type of universal joint, is found in airplanes and cars. This joint is a way of holding a linkage while allowing movement. In airplanes, ball joints are part of the control mechanism for the wing elevators. In cars, ball joints are part of the suspension system.

The ball joint is used in the construction of artificial knee and hip joints. Today, artificial knee and hip operations are routine, and engineers, doctors, and surgeons

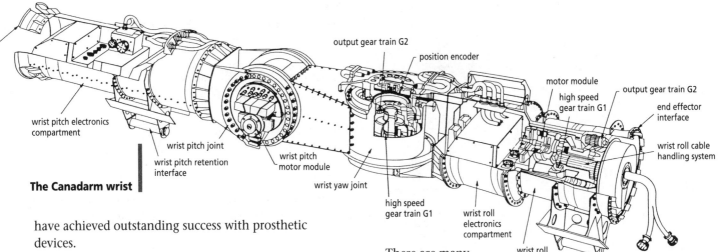

output gear train G2

position encoder

motor module

high speed
gear train G1

output gear train G2

end effector
interface

wrist pitch electronics
compartment

wrist pitch joint

wrist pitch retention
interface

wrist pitch
motor module

wrist yaw joint

high speed
gear train G1

wrist roll
electronics
compartment

wrist roll cable
handling system

wrist roll
joint

wrist roll retention
interface

The Canadarm wrist

have achieved outstanding success with prosthetic devices.

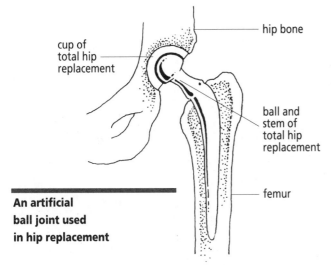

hip bone

cup of
total hip
replacement

ball and
stem of
total hip
replacement

femur

**An artificial
ball joint used
in hip replacement**

Using Energy to Power a Machine

*T*hrough the use of stored energy, you can make a machine work without having to be near it. As a designer, you should consider the use of energy when you are designing an object to fulfill a need.

For example, if you decide to design and fabricate a model airplane that flies, you may want an unwinding elastic band to turn the propeller. This would provide the force to keep the plane airborne. You would be using the energy stored in the elastic band to create the motion.

There are many types of energy that can be used to create motion in a machine that you design. Many are simple and will give you the freedom of not having to control your fabricated project directly. Consider the following sources of stored energy as possible additions to any machine you design.

Pneumatics and Hydraulics

Pneumatics uses compressed air or other gases to transfer motion. **Hydraulics** uses compressed water or other liquids to transfer motion.

Bicycle tire pumps use the principle of pneumatics. The pump action takes in air and then forces the air into the tire. A valve allows air movement in only one direction.

Excavating equipment uses the principle of hydraulics to move machine arms. Agricultural equipment uses hydraulics to lift and position cultivators. These systems, however, can cause serious damage if a leaking fluid catches fire.

 Caution: Extreme care must be taken around compressed air and hydraulic fluids. A serious accident could occur if the confined force is suddenly released.

See for Yourself

Discover how pneumatics works by attaching a plastic squeeze bottle to a balloon using a piece of plastic tubing. By squeezing the plastic bottle, you can make the balloon expand and contract. You can fix the balloon to a lever with tape or string to produce an action. You and a classmate can measure the increase in balloon size using a tape measure.

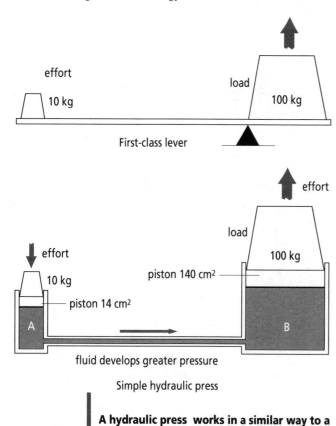

effort
10 kg

load
100 kg

First-class lever

effort

load
100 kg

effort
10 kg

piston 140 cm²

piston 14 cm²

A

B

fluid develops greater pressure

Simple hydraulic press

> **A hydraulic press works in a similar way to a simple lever. For a 10 kg effort to support a load of 100 kg, cylinder B must have a surface area 10 times that of cylinder A.**

Wind

The wind is a natural source of energy. A French space probe to Mars will use an expanding and contracting balloon to lift off the surface of the planet. Once in the air, the winds on the surface of Mars will move the balloon about. The balloon works on the principle that gas expands and rises when heated but contracts and sinks when cooled. The balloon will therefore rest at night, but will rise with the warming Sun and float around the planet during the day until the Sun's heat is gone.

Elastic Power

Elastic bands can store energy for release at a later time. The usual method is to keep the elastic twisted or stretched. In these forms, the elastic has potential energy. The energy is released when the elastic is allowed to untwist or relax. When this happens, the potential energy is converted to kinetic energy — the energy of motion. This energy is tranferred to a **propulsion** device — a device that propels or moves an object.

Suppose you are given the following need. As part of the design process, you would sketch some solutions.

EXAMPLE

Need: I need to design a toy that moves through the air on its own.

Brainstorming: (possible ideas)
- Create a paper airplane.
- Create a rocket.
- Create a space ship.
- Create an elastic-powered airplane.
- Create a hot-air balloon.

Design Brief: To design and fabricate an airplane propelled by an elastic band connected to a propeller.

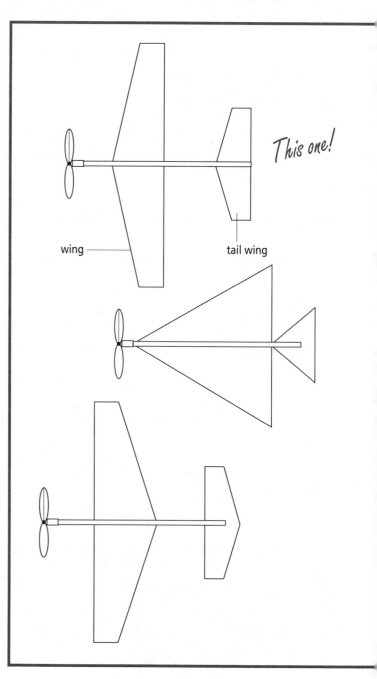

This one!

wing

tail wing

MAKING A PADDLE BOAT

Suppose you need to build a toy that moves in water. The toy is for a child. You want the child to be able to work this toy by him/herself. To approach this problem, you decide to use the design process described in Chapter 2. The following outline shows how this problem might be solved.

Need: I need to create a toy that will move through water.

Brainstorming: (possible solutions)
- Create a submarine.

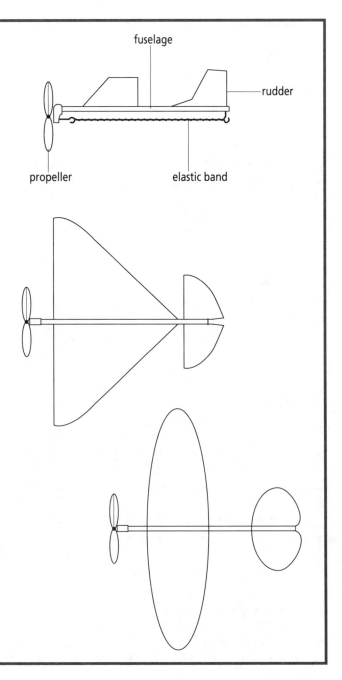

fuselage

rudder

propeller

elastic band

- Create a toy dolphin.
- Create a jet-propelled boat.
- Create a sailboat.
- Create an elastic-powered paddle boat.

Design Brief: To create a paddle boat that uses an elastic band to power the paddle.

Researching: I could do the following: Discover what makes boats move through the water. Talk to the science teacher. Check books in the library that explain transportation. Perhaps there are diagrams showing how old paddlewheelers were made. Maybe these would give me clues on what design to use. Consider the best shape and colour, the most suitable materials, and the most effective method of fabrication.

I should consider these questions: What material will be adequate? Will my boat be able to float? Will the paddle be able to push the boat through water? What type of elastic band should I use? Will my design and final project be safe for a child to use? How much time do I have?

I should draw up a decision-making chart to help me decide which design factors will be important. I will make a few sketches — at least five — with various shapes and measurements.

Planning: I will narrow down the choices and make a final selection. Perhaps I should make an exact-size model.

I should include a bill of materials. I will use plywood and a thick elastic band for the power. For waterproofing, I will use urethane. I will also need glue (probably epoxy) and measuring and cutting tools. A chart will help me plan the fabrication steps.

Fabrication: I will set up my tools in the order shown on my chart. I will be sure to check with my teacher for any particular safety precautions that I may have to consider when using the tools. I will fabricate the paddle boat using the exact measurements of my model and follow the procedures that I have recorded on my chart.

Evaluation: When the paddle boat is completed, I will conduct a series of test runs. In the evaluation, I will determine whether I have a good design, whether my paddle boat is able to move in the water as I had expected, and what modifications or improvements are required. I will ask myself honestly, "Did I fulfill my need?" In order that others may benefit from my research, planning, development, and testing, I will make a presentation of the stages of the project, the sketches, the problems encountered, the order of

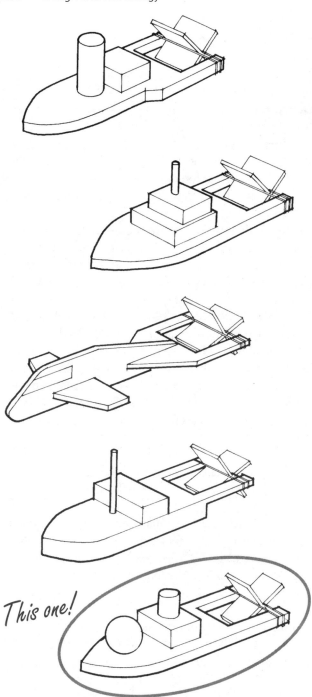

This one!

operations, the testing, and any changes made to the final project.

Springs

Like elastic bands, **springs** can be used to store energy. The spring in a wind-up motor can be wound and then allowed to release its stored energy slowly through a series of gears. Some wind-up clocks have gear mechanisms that can stretch the release of stored energy over

many days. A wind-up motor is an excellent device for making a project move. However, it needs rewinding at regular intervals.

Springs can be used to reverse motion. The expanded spring returns to its original position once the force has been expended.

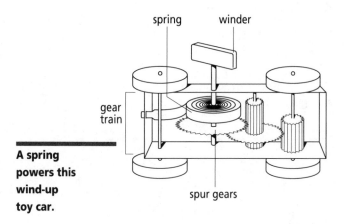

A spring powers this wind-up toy car.

Batteries

Using **batteries** as an energy source will allow you to use electric motors in your projects. You will need a 9 V or less direct current (DC) motor. Such motors are readily available in hobby shops. Their cost is minimal; you will probably spend more on the batteries to keep your project running.

An alternative to battery power is the use of a solar panel. Although you will need to buy the panel from a science or hobby store, the energy from the Sun is free. The solar panel is attached directly to an electric motor. When the panel is exposed to sunlight, the motor will move.

Elementary Robotics

Many of us are familiar with walking, talking robots made famous in movies and television shows. What exactly are robots? Can robots really walk, talk, and act like people, the way we see them portrayed in science fiction?

Real robots are simply motorized machines that are designed to do a variety of tasks. They are also a combination of several simple machines we have discussed so far: wheels and axles, pulleys, levers, gears, linkages, and joints.

A robot is composed of a mechanical unit and a processor unit. The mechanical unit is equipped with electrical, hydraulic, or pneumatic power that is controlled by a human. The human operator might be replaced by the processor unit, a computer that sends signals to the power controls.

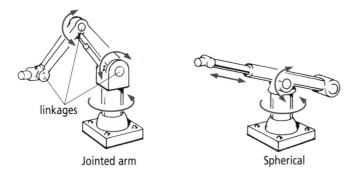

linkages

Jointed arm   Spherical

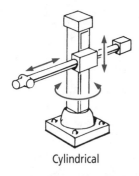

Cylindrical   Rectangular

Robots powered by electric motors are the most versatile and expensive. The type of motor determines how accurate and fast they are.

Robots have very limited degrees of movement. These are analyzed on the basis of the mathematical planes x, y, z, which describe the degree of robotic movement (see the diagram below). A robot has three identifiable components: (1) a wrist; (2) an arm made up of segments (rigid bodies that can move relative to the arm) and articulations (connections that limit the movement between the segments); and (3) the gripper. The robotic arm is similar in structure to the human arm, with a hand that has fingers and a thumb for grasping.

Types of robots

Unlike the human arm, the robotic arm has only five degrees of movement (such as gripping, lifting, and so on). The role of the robotic arm is to move the gripper ("hand") to a position where it can perform a task, such as spot welding metal panels. The stationary-arm robot used in industry (a robot that stays in one place) can perform a specific task all day long and not get tired. It can pile parts all day. It can do welding all day. Or, it can perform a task that a person might find boring or hazardous. However, robots cannot think. They must be programmed by a knowledgeable person. People need to develop these new skills, since robots are now doing tasks that people used to do.

Besides the stationary-arm robot, which is the standard in manufacturing, robots have been developed that move from place to place. For example, some police forces use robots to search areas for suspected bombs or to do other dangerous work. Robots with special sensors can be made to follow a line on a factory floor. Some companies use such robots to deliver mail within the office.

The benefits of "walking" robots are immense, especially in their contribution to space missions. With the development of high-power computers, more research into walking robots is being carried out.

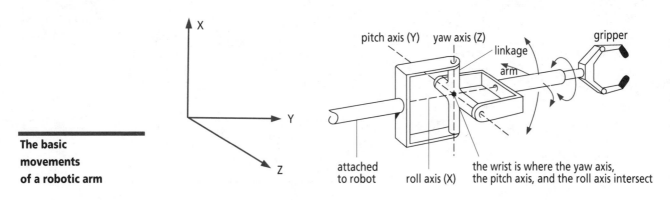

Robots such as these in an automotive plant are now commonplace in manufacturing.

The basic movements of a robotic arm

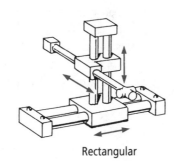

The study of mechanics has enabled us to understand how things work. It has given us the knowledge to design, fabricate, and operate machines. Machines will continue to be a major force in the development and progress of our world.

Points for Review

- Machines provide a mechanical advantage. Six simple machines that provide this advantage are the lever, the wheel and axle, the pulley, the inclined plane, the wedge, and the screw.
- A lever is a bar that magnifies the effect of a small effort to move a large load. The bar turns on a fixed point called the fulcrum.
- In a wheel and axle, effort applied to the wheel causes the axle to turn with a greater force than the wheel.
- A friction-reducing wheel does not transmit force.
- A pulley transmits force by means of a rope, belt, or chain passing over the rim of a wheel.
- An inclined plane provides a sloped surface to gain access to lower and higher levels.
- A wedge is an inclined plane that moves.
- A screw is an inclined plane wrapped around a cylinder.
- There are three principles of power transmission: a small driver wheel pushing a large driven wheel gives low speed and high force; a large driver wheel pushing a small driven wheel gives high speed and low force; a driver wheel and a driven wheel of the same size will have the same speed and force.
- A belt drive transmits power from the driver pulley to the driven pulley by means of a belt.
- A chain drive is similar to a belt drive except that toothed wheels and a chain are used, eliminating slippage and providing a more efficient means of transmitting force.
- Gears, which are wheels with teeth, transfer force from one part of a motor to another.
- Reciprocating drives change a rotary motion to a linear motion.
- Linkages consist of parts connected to one another by joints.
- Pneumatics uses air or other gases to transfer motion, whereas hydraulics uses compressed liquids, such as water or oil, to transfer motion.
- Elastic bands and springs are sources of stored energy.
- Robots are motorized machines that combine several simple machines. A robot is composed of a mechanical unit and a processor unit.

Terms to Remember

Archimedes' screw	effort	load	screw
auger	elastic bands	machines	second-class lever
bar linkages	first-class lever	mechanical advantage	shafts
batteries	follower rod	moment	springs
belt drive	friction	pantograph	third-class lever
brake calliper	fulcrum	pitch	torque
cable linkages	gears	pneumatics	transmit
cam	hydraulics	propulsion	universal joint
chain drive	inclined plane	prostheses	wedge
crank	lever	pulley	wheel and axle
crankarm and lever	linear motion	robots	wheels
crankshaft	linkages	rotary motion	worm gear
crankwheel and lever			

Applying Your Knowledge

1. The average design and technology classroom has many machines. Identify an example of each of the following machines in your classroom (remember, they may appear as parts of larger machines):
    a. a first-class lever,
    b. a second-class lever,
    c. a third-class lever,
    d. a wheel and axle,
    e. a wedge,
    f. a screw,
    g. an inclined plane.
    Compare lists with a classmate. Are there any differences in your lists? Double-check each other's lists to ensure that the correct machines have been identified.
2. Label and sketch the principal shapes of
    a. a lever,
    b. an inclined plane,
    c. a wedge,
    d. a wheel and axle,
    e. a pulley.
3. Sketch a tennis racket and label the following: load, fulcrum, and effort. Is the tennis racket a first-, a second-, or a third-class lever?
4. A driver wheel transfers force to a driven wheel. In three short sentences, state the three rules of wheel size and wheel direction.
5. A small grocery store has very high shelves. Using the design process, design a device the store owner can use to reach the cereal boxes on a top shelf. (HINT: You can use any of the simple machines discussed in this chapter or any combination of them.)
6. Examine the following diagrams:

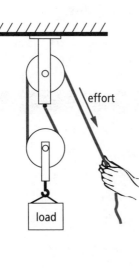

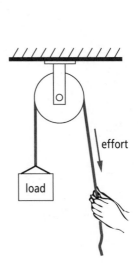

a. What are these wheeled devices called?

b. Describe briefly how the two systems differ.

c. Explain the mechanical advantage of pulley A and pulley B.

7. Imagine that you have fabricated a toy car. List ways that you could use stored energy to make the car travel across the floor.

8. Make a list of all the wheels that are found in your home. Divide this list into two lists: wheels that aid directly in transportation, and wheels that reduce friction.

9. Examine the following diagram:

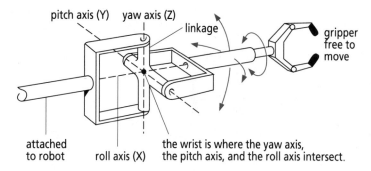

pitch axis (Y) yaw axis (Z) linkage gripper free to move

attached to robot roll axis (X) the wrist is where the yaw axis, the pitch axis, and the roll axis intersect.

a. In Chapter 4, the words pitch, yaw, and roll were used to explain a seemingly unrelated topic. What was this topic?

b. Explain in two or three sentences the connection between your answer above and robotics, with respect to pitch, yaw, and roll.

10. Imagine yourself in the following role: You are about to open your own manufacturing company to make anything you want. Make a list of the pros and cons of using robots in your company. Share your ideas in a group discussion on whether robots should be used in manufacturing. Present your group's list to the class to make a master list of the pros and cons of using robots.

11. How is a robotic arm similar to and different from a human arm? Make a chart listing these similarities and differences. Compare charts with a classmate.

12. How have the developments in robotics and computers affected the types of skills that people need for the job market?

Chapter 8

MATERIALS

What You Will Discover

After completing this chapter, you should be able to:

- Identify the wide range of products in our society that are made of woods, plastics, metals, and ceramics.
- Test materials and select the best one to suit your design needs.
- Understand the reasons for selecting certain materials.
- Understand the importance of recycling materials and why recycling is becoming more popular.

*I*n ancient civilizations, people's needs were few and simple. When people faced a need, they used the materials around them, exactly as they were found. Leaves and branches were used to build shelters, and pieces of stone and wood were used as containers.

As civilization progressed, people's needs increased and changed, and new **materials** were discovered or developed to meet these demands. Today, there is a large range of materials available to designers, craftspeople, and manufacturers.

As a student, you need materials to fabricate the designs that you create. It is important for you to know the kinds of materials that are available, the sources of these materials, and their characteristics. You also need to be aware of the kinds of materials that are environmentally friendly and that can be recycled or reused.

Materials and the Design Process

*L*ook around you — every manufactured product you see has been made using one, two, or many materials. Fulfilling a need through the design process means that you will have to consider which materials to use. This occurs in the researching stage. There are special reasons why materials must be carefully considered before you begin to fabricate a project.

In most cases, you will be designing something that has **physical qualities**. This means that your fabricated project will have a particular shape and structure.

The shape and structure of your project will be determined by the materials you select. The materials will also determine how your project functions. (For more on shape, see Chapter 3; for more on structure, see Chapter 5).

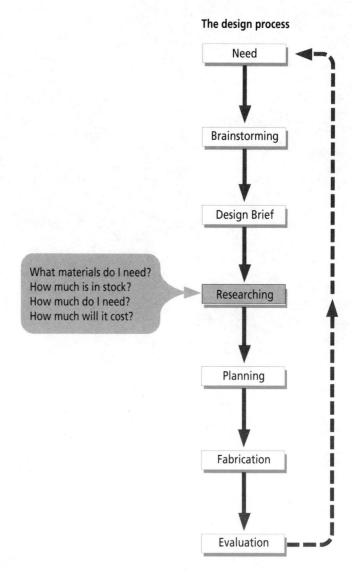

The design process

What materials do I need?
How much is in stock?
How much do I need?
How much will it cost?

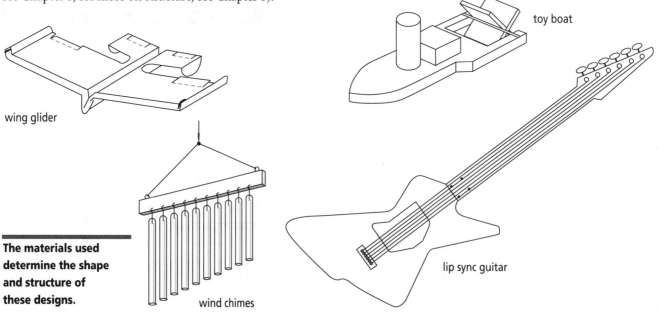

wing glider

toy boat

lip sync guitar

wind chimes

The materials used determine the shape and structure of these designs.

Materials can be evaluated on their aesthetic qualities: how they look, feel, and smell. These qualities cannot be measured exactly — different qualities appeal to our emotions or tastes. To evaluate the aesthetic appeal of a material, you may ask yourself some of the following questions:

- How does the material appear? If it is wood, does it have an attractive grain? If it is metal, is it speckled? What colour is the material?
- Is the material smooth or rough? Is it cold or warm to the touch?
- Does the material have a particular smell? Different woods have different smells. Cedar, for instance, has a distinctive smell.

Use and Waste of Materials

A constant concern of people who are involved in design and technology is using materials with as little waste as possible. To understand why reducing waste is so important, it helps to look at "the small picture" and "the big picture."

The Blue Box program is a response to material waste.

Take a look around your classroom. Try to count all the materials you see. Each material has a particular cost. Your teacher orders certain amounts of each material based on expected use. Now imagine that your classroom is a small business and that you are the owner. You must make the most of all the materials you order. When you waste a material, you are also wasting the money used to buy that material. When you are working with materials in the classroom, keep in mind the following:

- Choose a piece of material that is close to the size to be cut.
- Mark on the material exactly what you need for your project.
- Check with your teacher before you cut any material to ensure that your measurements are correct and you are not wasting material.
- Do not cut a piece from the centre of a material.

- Place leftover pieces in a storage container. These pieces may be used in other projects or by other students.

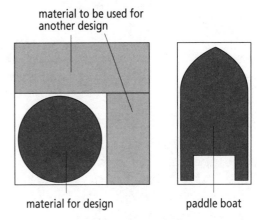

material to be used for another design

material for design   paddle boat

You should always try to waste as little material as possible. Your fabrication should be cost-effective and should conserve natural resources. Suppose that, on average, you are wasting one-third of all the materials you use. That means each time you order new material from your supplier — for example, wood from a local lumberyard — you are ordering one-third too much. To put it another way, one-third of all the trees it takes to supply your order are being wasted. Now suppose that all other users are like you, wasting one-third of their wood. This means that one-third of all trees harvested are being thrown away. It takes many years for the same amount of trees to grow to replace the ones wasted.

Try to cut material so that there is little waste.

Suppose you always waste one-third of your materials. What are the long-term effects of this waste?

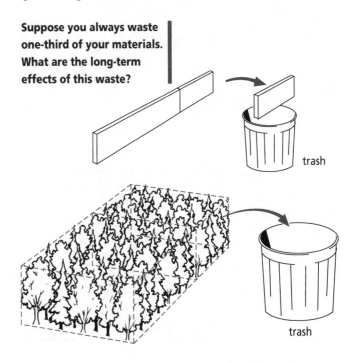

trash

trash

Career Profile

Heidi Overhill

"Design today is finally attracting popular attention, now that we realize how important it is to the success of every product, and to the success or failure of the manufacturing company," says designer and entrepreneur Heidi Overhill. "This is an exciting time to be in our profession."

Like many other successful industrial designers, Overhill's first interest was art. She has been able to use this interest in industrial design, which she describes as "art, social sciences, and engineering all rolled up into one fascinating career package."

"After graduating with distinction with a Bachelor of Industrial Design from Carleton University, I still felt there was more to learn, particularly about the history of design," says Overhill, "so I applied for and won a scholarship for post-graduate studies in England, at the Royal College of Art. Afterwards, I worked briefly in Italy as an unpaid apprentice product designer before returning home to Canada."

Later, Overhill worked extensively in museum exhibit design. This involved planning the interior wall construction for new galleries, as well as designing display cases, labels, signs, and information sheets. She also supervised the suppliers and builders to ensure that all work was completed satisfactorily and on time. Eventually, Overhill became senior designer at the National Gallery in Ottawa.

Now, as an entrepreneur, Overhill is a consultant to various clients. She also runs a small manufacturing company that makes quilts and other textile products in India.

Heidi Overhill and some of her quilt designs

QUESTIONS AND EXERCISES

1. What are Heidi Overhill's responsibilities when designing museum exhibits?
2. Choose a textile product that you would like to design. Make several sketches of possible designs.

Recycling

Most materials come from non-renewable resources — that is, they will be exhausted one day. Even wood takes many years to replace. It is important to recycle materials such as glass, metal, wood, paper, and plastic. In your classroom, you should have designated containers for scrap materials to be reused and materials to be recycled. Be sure you know where these containers are, and use them.

We have a responsibility, not just to ourselves but to society as well, to minimize waste. If we do so, we will be getting the most out of our resources.

Material Selection: Where to Start

*Y*ou should reach the researching stage in the design process before selecting your materials. This means you will have already completed the following steps:
- discovered the need,
- brainstormed possible solutions, and
- completed a design brief.

Your research will probably give you some clues as to what materials to consider. Considering the design factors will help you decide which materials to choose. It will show you the important qualities, in terms of materials, that your final project should have. The fulfillment of these qualities will be achieved only by selecting the best materials for the purpose.

The questions you should ask yourself when choosing materials are:
- What materials are available (in stock)?
- Which materials will best suit my design?
- What will be the cost of the materials I select (what do I need to buy and how much will it cost)?
- What materials are environmentally friendly and can be recycled?

General Categories of Materials

In our modern society, an immense range of materials is available. For simplicity, this textbook assumes that woods, plastics, metals, and ceramics are the standard materials that are available to you. In this chapter, you will learn how to go about choosing between these four materials. The same method can be used to decide which type of wood, plastic, metal, or ceramic would be best for your project.

You may need another common material for a project, such as paper. If you want to use a less common material, be sure to consult your teacher before doing so.

Material Properties

Once you know what materials are available, you can begin to select the best ones for your project. You should always select materials based on your needs. You may need more than one material; for example, a lamp may require a metal stand and a plastic shade.

All materials possess certain **properties**. That is, they have observable qualities over a period of time. Materials that are used in fabrication usually possess some or most of the following properties:

- Hardness — resistance to scratching, wear, and strong impact.
- Mass — how heavy the material is.
- Water Resistance — how well the material resists water damage.
- Heat Resistance — how well the material resists heat damage.
- Resistance to Fatigue — resistance to weakening caused by overuse or strain.
- Malleability — the ability to be beaten and rolled without breaking.
- Elasticity — the ability to return to the original shape after having been bent, twisted, or deformed.
- Ductility — the ability to be stretched or hammered into fine threads or wires.
- Brittleness — the property that makes a material shatter or break up.
- Toughness — resistance to fracturing or breaking.
- Tenacity — resistance to fracturing when stretched, measured in tensile strength.

Professional designers consider these properties of materials in order to decide which material is best suited to a product. A material will be tested for the desired properties and rated on its performance. You will likely consider these common properties at one time or another. The aim is to understand what properties are needed in a material, and to know how to select the best material.

Testing Material Properties

The only way to ensure that your design needs are met is to know which material best fulfills the properties demanded by your design.

The properties of most common materials are known, since a great deal of testing has been done on them. This sophisticated testing takes place in laboratories of research companies. Often, however, you may have to perform your own tests to determine how materials compare on a certain property.

Rating Some Properties of Woods, Plastics, Metals, and Ceramics

If you were to perform your own tests on woods, plastics, metals, and ceramics, you would be left with ratings of each of their properties. It would be important to **standardize** your tests. This means that, for each property, you would use the exact same testing procedures (such as in the See for Yourself on the following page). You would also use the same size and shape of each piece of material.

For example, suppose you need to fabricate the toy boat discussed in Chapter 7 (see page 119). Since you want your boat to float, a low mass will be an important property. You could select samples of different materials of equal size and place them in a pan of water. You would then observe which material displaces the least amount of water. Or, you could weigh the materials on a scale. Your tests show that a certain type of wood (for example, basswood) is the lightest material. You should probably consider using this material first.

Knowing the properties of available materials can save you a lot of time. You can develop quick-reference charts to remind you of the important properties of these materials based on the tests you do and the experiences you have with different materials in different projects.

Materials	The Factors that Affect Design										
RATE: 5 = excellent 4 3 2 1 = poor IDEAS: (Alternatives)	Safety	Function	Aesthetics	Ergonomics	Time	Materials	Environment	Hold	Reject	Modify	TOTAL
Plastic Boat	5	4	4	4	5	4	3			X	29
Metal (Tin) Boat	5	2	3	3	3	4	3		X		23
Wood Boat (with urethane coat)	5	5	5	5	4	5	4	X			33
Ceramic (clay) Boat	5	1	2	1	2	2	5		X		18

Rating materials for a toy boat

Discover the hardest material in your classroom. Select a small sample of each of four different materials that are readily available. Place one of the samples on the floor. Hold a 1 m length of metal tubing upright, so that one end rests on the sample. Insert a centre punch in the end of the tube, with the pointed end facing the sample. Let the centre punch fall. Examine the sample for the indentation. Repeat this test on the remaining samples. Which material has the deepest indentation and is therefore the least hard? Which material is the most hard? Rate the materials from 1 (most hard) to 4 (least hard). Discuss your findings with the rest of the class.

Design a test to compare the water resistance, heat resistance, or mass of four materials.

Choosing Your Material

When you have decided which material you think is best, you should discuss it with your teacher. Remember, this is a process of **consultation** — that is, you are asking for advice. Consultation is an activity that designers never outgrow.

You and your teacher should discuss your material choice based on the needs of your design, how the material can be fabricated using the technology available (for more on this topic, see Chapter 9), the cost of the material, and its effect on the environment.

The cost is affected by the size and amount of material needed. Sometimes you will have to reject the best material because it is too expensive. You will then have to choose an alternative material.

You and your teacher should also discuss your material choice in terms of how it affects the environment. Ensure that you are not using rare materials without considering what effects this may have. For example, mahogany is a rare wood that grows in rain forests. Buying this wood encourages the cutting of this tree, which damages the rain forest. Also consider how easily a material can be recycled.

Narrowing Down Your Choice

Choosing the general category of material — woods, plastics, metals, or ceramics — is a very important start. However, once you have done this, you must ask yourself, "Which specific type should I use?" For example, you may have decided to use wood as your main material, but which type of wood? Usually there are several types from which to choose.

The following sections give you information on specific woods, plastics, metals, and ceramics. You will also learn how to select the best of each type.

Woods

*W*ood is a very popular medium to work with for a variety of reasons. It can be easily shaped with hand or power tools. It can be cut to different thicknesses and sizes. Depending on the grain structure, it is strong and can support a fair mass. Wood can be filed and sanded to a pleasing finish. It also has the ability to absorb shock, and it provides good insulation against heat and sound.

The Carmanah Valley in British Columbia. Logged areas of mature forests must be replanted to ensure that there will be wood for the future.

Wood is one of our most abundant natural resources. The two main categories of wood are softwood and hardwood. **Softwood** usually comes from evergreen, or coniferous, trees. These are trees that keep their needlelike leaves all year round. Examples are pine, cedar, spruce, cypress, fir, hemlock, juniper, larch, redwood, tamarack, and yew. **Hardwood** usually comes from deciduous trees, whose leaves are broad and flat and fall off in autumn. Examples are birch, maple, oak, ash, poplar, basswood, and walnut.

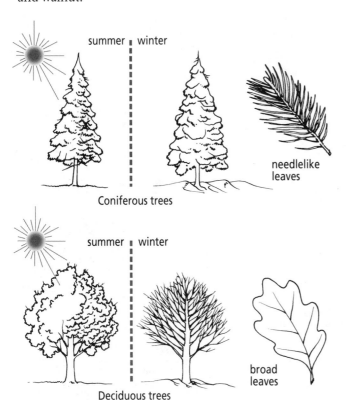

How to recognize coniferous and deciduous trees

Coniferous trees

Deciduous trees

Wood is a renewable resource. By planting new trees and caring for our forests, we can maintain an ample supply of wood to meet our needs. However, more trees are usually cut than are planted. Even though wood is so abundant, we must remember that it takes at least twenty to fifty years for trees to grow to a size suitable for harvesting.

Structure of Wood

One of the great advantages of wood is that its natural structure does not need to be altered much before it can be used in fabrication. The wood's natural structure provides strength, durability, and flexibility.

A cross section of a tree trunk shows the main parts of the tree.

- **Bark** — This is the outer, protective coating of the tree. Without bark, the tree cannot survive.
- **Cambium** — This is the layer where the growth of the tree actually takes place.
- **Annual rings** — As the tree grows, the cells containing sap or moisture develop along the cambium layer of the tree. With the abundance of moisture in spring, rapid growth takes place. This produces a light, wide layer of wood called the spring growth. As the season progresses, there is less moisture and slower growth, resulting in a darker, narrower layer of wood called the summer growth. When the tree is cut, these layers can be identified as annual rings. You can find the age of the tree by counting the number of rings.
- **Heartwood** — This is the inner part of the tree.
- **Sapwood** — This is the outer part of the tree. It is lighter in colour and softer than the heartwood.
- **Grain** — The grain indicates the degrees of contrast between early wood (spring growth) and late wood (summer growth). It is produced by the lengthwise cells in the tree. The grain is the pattern on the face of a board.

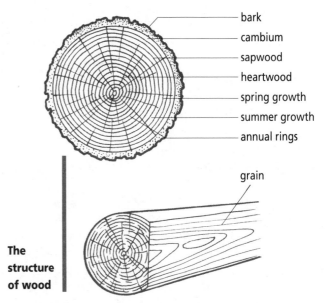

The structure of wood

How Lumber Is Processed

Trees are harvested and processed in a way that uses this wood structure to its maximum potential. The main form of processing wood is to turn it into lumber.

CUTTING

After trees are cut into logs, the logs are cut into boards called **lumber** in a sawmill. Lumber comes in various lengths, widths, and thicknesses. The three methods of cutting logs are slash (plain) cutting, quarter cutting, and rift cutting. These terms indicate the angle of the grain in relation to the saw. The method used is determined by the type of wood (softwood or hardwood) and the purpose for which it is intended (furniture making or construction work).

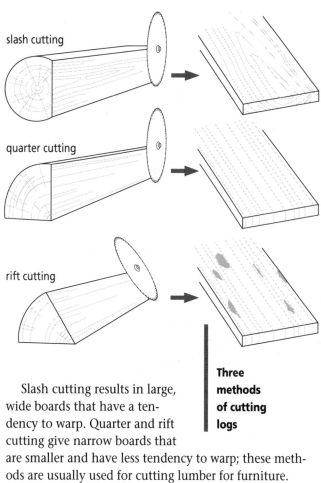

slash cutting

quarter cutting

rift cutting

Slash cutting results in large, wide boards that have a tendency to warp. Quarter and rift cutting give narrow boards that are smaller and have less tendency to warp; these methods are usually used for cutting lumber for furniture.

Three methods of cutting logs

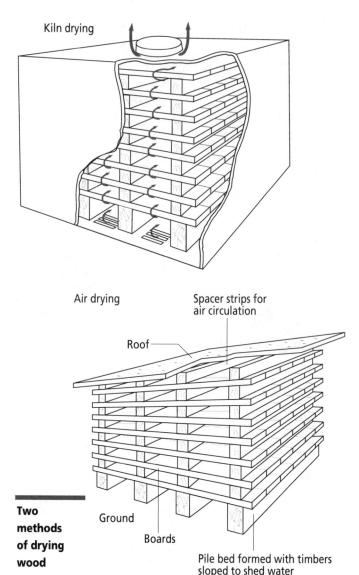

Kiln drying

Air drying

Spacer strips for air circulation

Roof

Two methods of drying wood

Ground

Boards

Pile bed formed with timbers sloped to shed water

DRYING

When logs are cut, there is an excessive amount of water in the lumber. The amount or percentage of water found in lumber is called its **moisture content**. The moisture content must be reduced before working with the wood. The lumber must be dried, just as clothes that are soaked with water must be dried either outside on a clothesline or in a dryer. Lumber can be dried outside (air drying) or in a dryer (kiln drying or radio-frequency dielectric drying).

During the process of drying, some warpage, or twisting, may occur. Warpage occurs because the cells of the wood dry at different rates when moisture is removed during the drying process. Shrinkage occurs most in the direction of the growth rings (width of the board) and very little across them (length of the board). Cracking occurs when wood dries too quickly. Controlling the drying process helps to minimize any warping, shrinking, and cracking of the lumber.

MEASURING

There are two predominant systems of measures in the world today: the imperial system (using feet and inches) and the metric system (using metres and millimetres). A number of years ago, Canada converted from the imperial system to the metric system. At present, there is a mix of imperial and metric measurements of lumber — a sign of the slow change from the old system of measurement (imperial) to the new system (metric). In a lumberyard, wood is usually sold by the linear foot, which is the length of a board. The linear foot does not take into account the thickness or width of a board. Wood is also processed into sheets measuring 1.2 m by 2.4 m (4 feet by 8 feet) with varying thicknesses from approximately 1.5 mm to 51 mm (1/16 to 2 inches).

DRESSING

Natural lumber from the sawmill has a rough surface. Usually it is smoothed on both sides using a machine

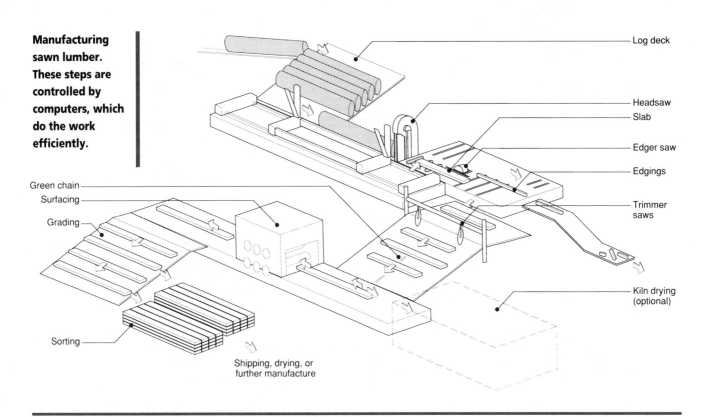

Manufacturing sawn lumber. These steps are controlled by computers, which do the work efficiently.

Log deck

Headsaw

Slab

Edger saw

Edgings

Trimmer saws

Green chain

Surfacing

Grading

Sorting

Shipping, drying, or further manufacture

Kiln drying (optional)

called a plane. This form of preparation is called **dressing**.

Lumber is commercially sawn and dressed to standard thicknesses of 6, 10, 13, 18, and 19 mm (1/4, 3/8, 1/2, 11/16, and 3/4 inches) and to standard widths from 5 to 30.5 cm (2 to 12 inches), progressing in intervals of 2.5 cm (1 inch) and to standard lengths of 2.4, 3, 3.7, 4.3, and 5.5 m (8, 10, 12, 14, 16, and 18 feet) or longer. However, lumber can be dressed to any desired thickness and cut to any desired length.

WOOD DEFECTS

Defects reduce the strength and desirable appearance of wood. You should check lumber for defects before using it.

One defect is a knot: a compact, hard, cross-grained mass that is the starting point of a limb or branch. Some knots are loose, while others are solidly embedded in a board. Sometimes knots are desirable — for example, in knotty pine furniture.

A warp, or bend, is caused when wood is dried improperly or too quickly.

A check is a split in the wood that is parallel to the grain. It is caused by stresses during drying.

A worm hole is a small, pinlike hole caused by termites or insects.

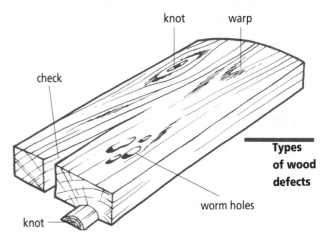

knot

warp

check

worm holes

knot

Types of wood defects

Recognizing and Choosing Woods

Suppose you have gone through your preliminary decision making and have decided that the use of wood would best fulfill your design requirements. In your classroom, you probably have several types of wood, so you will have to make a specific choice.

All woods have a recognizable colour and grain. You will be able to tell one wood from another by referring to the following chart.

Woods

Name	Colour	Grain	Properties
Basswood	very light brown heartwood, nearly white sapwood	fine, even texture and grain	exceptional stability, good machining and sanding characteristics
Birch	pale brown	medium texture, straight grain	hard, versatile finish
Hard maple	very light brown or tan heartwood, white or off-white sapwood	very fine texture and grain	hard, heavy, strong, resistant to abrasive wear
Red cedar	red hardwood, white sapwood	irregular, knotty, close grain	good machining properties, non-porous, durable
Spruce	pale cream	fine grain	works easily, holds fasteners well
White ash	white to light tan	coarse texture, straight grain	heavy, tough, hard, high shock resistance, machines well
White oak	pale tan	medium texture; straight, close grain	hard, machines well
White pine	cream to light tan heartwood	fine texture; soft, uniform grain	very responsive to carving tools, takes intricate detail cleanly

Although your own situation may differ, most design and technology classrooms carry birch, spruce, pine, and basswood. An example will show how you might go about selecting the best of these four woods.

Rating types of wood

	Poor ⟶ Excellent				
		1	2	3	4	5
Mass of wood			*birch*	*spruce*	*pine*	*basswood*

Example: Wood for a Toy Boat

In Chapter 7 you were given the example of making a toy boat (see page 119). Suppose you are deciding among birch, spruce, pine, and basswood for the boat. You must remind yourself of the important design factors.

Because you want the boat to be able to float in water, it must be light and water-resistant. Since you decide that any wood can be covered with a coat of urethane, water resistance is no longer a major concern. That means you must test the wood samples to find the lightest of the four.

You can see from the chart that basswood is the lightest wood, followed by pine. You decide that basswood is the best choice for your toy boat.

Consulting Your Teacher

As part of the design process, you should consult your teacher before you begin fabrication. Material selection is part of researching and is one of several important decisions you must discuss.

See for Yourself

To discover which wood in your classroom is the lightest, cut equal-sized cubes of each type. Then fill a tub or sink with water, and put the samples in the water. Make sure that the water is calm. Observe how far down the wood samples sit in the water. Which cube has the greatest mass? Which has the least mass? Rank the masses from greatest to least. Discuss your observations with a classmate.

Discussing the best material to use will also help you to decide on the fabrication steps. In some cases, your choice of material will be affected by the available means of fabrication. For instance, you may have chosen the best material but may have no available means of joining it. (For more details on fabrication, see Chapter 9.)

You will also consider environmental concerns. Are you using a material that is in rare supply? Will you be able to recycle any leftover material?

In the toy boat example, there may have been some important factors of which you were unaware. For instance, you may discover that the stock of basswood is low and that this type of wood is very expensive and not the proper size. In contrast, there is plenty of pine in stock, and pine is relatively inexpensive. This would mean that pine is the best choice.

Types of Manufactured Woods

*B*irch, spruce, pine, and basswood are common natural woods. There are other, manufactured woods with which you should be familiar.

Plywood

Plywood is a board made of several thin layers of wood glued together. Each layer is called a **ply**. The plies of wood are placed on top of one another with the grain of each ply at right angles to the one below it. Each layer is glued to the previous layer to make up the required thickness, and then the layers are put under heavy pressure, machine-sanded to the final panel thickness, and trimmed to the final dimensions.

The top and bottom layers of plywood are usually veneer. **Veneer** is a very thin layer of wood, similar in thickness to bristol board, that is usually of a superior quality.

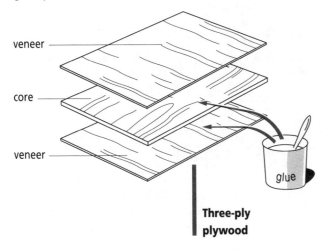

Three-ply plywood

Sheets of plywood are very strong, and they make durable and attractive panelling. The face veneer can be any type of wood that is used in cabinetmaking.

Hardboard

Hardboard is made of wood fibres that are bonded with an adhesive and subjected to great pressure at very high temperatures. This makes a tough, dense sheet that is available in various thicknesses.

A wide variety of hardboards are moulded or embossed, or surfaced with enamel, melamine, plastic, or veneer. These materials are used for panelling, siding, trays, tabletops, store fixtures, and backs of cupboards.

Particle Board

Particle board is a hard-surface board made of wood chips, splinters, flakes, and screened sawdust (mill wastes). These are finely milled to a uniform size and bonded with a formaldehyde glue for indoor applications. They are bonded with phenolic resin for exterior use to be more resistant to moisture and heat. Examples of particle boards are chipboard, flakeboard, and splinterboard.

Particle board is used for shelving, cupboards, and partitions. It is also used as the foundation for laminate-covered countertops and furniture. It is less expensive than plywood and hardboard of comparable thickness and comes in sizes similar to plywood.

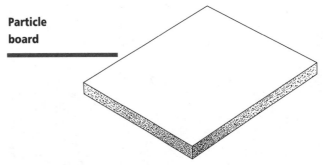

Particle board

Plastics

*T*he plastics industry is growing at a faster rate than most other manufacturing sectors in this country. **Plastics** are synthetic materials that are composed of the hydrocarbons found in petroleum. Some of these hydrocarbons are methane, propane, butane, and ethane.

The plastics industry started in North America in 1870, when the American scientist John Hyatt invented the first commercial plastic, called celluloid. This new material was made to replace ivory in billiard balls.

Career Profile

Susan Ramsay

A good example of how a need has been solved with design is the MOVETECH® moving box, created by Susan Ramsay. It all happened when Susan was planning to move — all she could find was an old, damp cardboard box. She felt that there must be a better way to move, and she knew that she was not the only person to face problems in moving.

That's why Ramsay started The Moving Store, where all the necessary supplies for moving could be found in one place. At first she stocked plastic boxes that the automotive industry uses to ship parts around Canada. But Susan soon developed a wish list for a true moving box. The box had to be

- easily stacked, carried packed, and cleaned, and
- fire-retardant, waterproof, dustproof, secure, and durable.

After years of research, Ramsay fabricated and patented her revolutionary plastic container called the MOVETECH® box. The success of The Moving Store — now a million-dollar business — is partly due to the success of the MOVETECH® box, since Ramsay obtained sole patent for it. Even if the box is broken, the material can be recycled.

QUESTIONS AND EXERCISES

1. Why did Susan Ramsay create the MOVETECH® box ?
2. List the qualities of a design that you want to fabricate.

Plastics are increasingly taking the place of woods, metals, and ceramics because they are inexpensive and easy to mould. In fact, plastics possess structural advantages over the other materials. Plastics are light yet strong and can be flexible or rigid. They can be machine formed, finished, and coloured. Plastics resist rotting and corrosion and are effective electrical and thermal insulators.

These awards are made of acrylic plastic.

These binders are made of vinyl plastic.

This children's ride has a plastic body.

These containers are made of polyethylene plastic.

Plastics are used in the manufacture of many products in place of other, traditional materials. You will find plastics in housewares, appliances, toys, furniture, sports equipment, jewellery, paints, car parts, building materials, aerospace parts, and prosthetic devices.

How Plastics Are Processed

The processing of most plastics begins with the refining of crude oil or coal. The substance called plastic is created through the various stages of heating and adding other chemicals.

Manufacturers use chemical reactions to change plastic bits called "feedstock" into plastic resins. These resins are then turned into forms that are useful for processing, such as granules, powders, pellets, flakes, and liquids. These forms of plastic are then sold to other manufacturers, who process them into plastic products by applying heat to melt them and pressure to force them into the desired shapes. Then the plastics are allowed to cool and set into the finished products.

Moulding, compressing, and forming are examples of the kinds of processing that plastics undergo to become final fabricated products. These are also three ways that you may process plastic while working on a design project. (For more on processing plastics, see Chapter 9,).

Environmental Concerns and Recycling

The use of plastics has raised some environmental concerns. Some types of plastic manufacturing cause air and water pollution. Many plastics also give off noxious fumes when they burn. Certain types of foam plastics rely on the use of chlorofluorocarbons (CFCs), which are a major cause of the deterioration of the Earth's ozone layer. The ozone layer shields the Earth from much of the harmful radiation from the Sun, so it must be protected. Many industries that formerly used CFCs — for example, in building insulation, fast-food containers, and packing materials — are now turning to more environmentally friendly products.

Programs for recycling plastics (as well as other materials) are rapidly gaining popularity and acceptance. Plastics can be recycled by grinding them down to pellets and feeding them back into a processing machine to make other products.

Recognizing and Choosing Plastics

*P*lastics have many desirable properties that might fulfill some of your design needs. In many ways, plastic combines the properties of wood and metal. Like wood, plastic is generally light; like metal, it can be bent and formed into shapes.

There are many types of plastics used in design. They are classified in terms of their reaction to heat. Plastics fall into two general categories: thermoplastics and thermosets.

Thermoplastics

Plastics that liquefy when they are heated and solidify when they are cooled are **thermoplastics**. They are purchased as pellets or granules and are then softened by heat under pressure so that they can be formed and then cooled to harden into the desired shapes. The process is similar to melting a block of ice, pouring the water into any shape of container, and then freezing it into a solid again. The thermoplastic process can be repeated indefinitely.

Examples of thermoplastic products are plastic bags, soft drink and detergent bottles, water pipes, hoses, raincoats, packing materials, and ceiling tiles. Thermoplastics that you are likely to use in your class are **acrylic**, **polyvinyl chloride (PVC)**, **polyethylene**, and **acrylonitrile butadiene styrene (ABS)**.

A very important characteristic of acrylic is its outstanding resistance to long-term exposure to sunlight and weathering. Acrylics are strong, rigid, and more impact-resistant than glass. They are often used as light diffusers. You will find acrylics in aircraft canopies and windows, outdoor signs, skylights, car tail-lights, room dividers, and home furnishings such as tables and lamps.

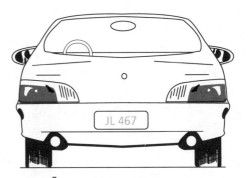

The tail-lights of this car are made of acrylic.

PVC is used in upholstery, flooring, wire insulation, hoses, siding, pipes, credit cards, boots, and many other items.

Polyethylene is the most widely used plastic. It is lightweight and strong. Polyethylene is made into garbage bags, toys, housewares, containers, stadium seats, and automotive parts.

ABS plastics that come in sheet form can be vacuum formed, pressure formed, or cold stamped. They are made into telephones, safety helmets, luggage, and car parts.

Nylon is another thermoplastic. It is extremely strong and durable and has many uses, especially in textiles, kitchen utensils, and brush bristles.

Thermosets

Plastics that solidify under heat and cannot be made into a liquid again are **thermosets**. They can be formed into finished shapes by heating or by chemical processes. Examples of thermoset products include boat and car bodies; repair kits for adding fibreglass patches to a boat, car, or snowmobile; tableware; and surfaces for counters, tables, and cabinets. Some thermoset products, such as clear cast ornaments, require special mixing and clean-up procedures, which limit their use in the classroom.

Epoxy resin is an excellent thermoset plastic for bonding. It has great strength and superior adhesive characteristics as a glue for bonding metals, plastics, ceramics, glass, and hard rubber.

Polyurethane is a widely used foam plastic. This thermoset is extremely lightweight and is used in car parts and as a packing material.

Although two plastics may perform the same function, they may be available in very different forms. The chart below lists the basic plastics and describes their properties.

The dashboard and steering wheel of this car are made of polyurethane plastic.

Example: Plastic for an Organizer

Suppose that you have decided to design and fabricate a plastic organizer to hold mail, pencils, and pens. You will have to decide what type of plastic to use. The most important design factors are
- the ability to be bent and formed, and
- rigidity, to support what is placed in the holder.

Using the chart of plastic properties, you decide to use acrylic to make your organizer. This plastic is easy to bend when heated and cools to a rigid form.

Plastics

Name	Colour/opacity	Properties	Handling	Common uses
ABS	colour varies, opaque	semi-rigid, sinks in water	cuts smoothly, burns easily	refrigerator liners, cases for electrical machines
Acrylic	colour varies, clear to opaque	rigid, sinks in water	chips when cut, burns easily	kitchen utensils, housewares, table tops, furniture, car parts
Epoxy resin	colour varies, translucent to opaque	rigid, sinks in water	chips when cut, burns easily	glue, paint
Nylon	colour varies, opaque	rigid, sinks in water	cuts smoothly, does not burn easily	gears, bearings, curtain rod fittings
Polyethylene	colour varies, clear to opaque	soft to semi-rigid, depending on type; floats in water	cuts smoothly, burns easily	pipes, squeeze bottles
Polyurethane	any colour, opaque	soft to rigid, depending on type; floats in water	cuts smoothly or chips easily, depending on type; burns easily	car parts, seat cushions
PVC	colour varies, clear to opaque	semi-rigid, sinks in water	cuts smoothly, does not burn easily	wire insulation, floor coverings, clothing

Consulting Your Teacher

You should consult your teacher to discuss your choice of plastic. Your teacher may tell you that acrylic is a good choice for your organizer because it has the properties you need and comes in a variety of colours.

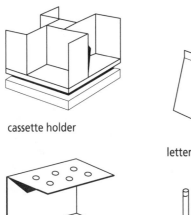

cassette holder

letter or napkin holder

pencil holder

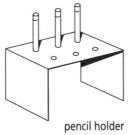

pencil holder

Your plastic organizer can be a combination of these common holders.

Two uses for the same metal: copper kettles and copper anodes for batteries

How Metals Are Processed

The structure of metal is different from other materials in that it can withstand great heat but still hold together. This allows it to be processed in different ways from other materials.

For example, wood will burn when heated beyond a certain point. The resulting material cannot be processed to make solid wood again. Metal can be heated to another state — liquid metal — and modified before it is cooled back to its solid state.

SMELTING

Most metals are found in an impure form as metallic minerals in rocks. In this form, the metals are called **ores**. Ores are mined, broken up, and smelted in order to obtain the purest quality metals. This means that they are a non-renewable resource — they cannot be replaced.

Metals

Metals have been used for thousands of years in many ways. They replaced wood in many uses, because metals have many advantages over woods. Metals can outperform woods in almost all the material properties, and they also have some properties that other materials do not. For example, metals conduct electricity and heat. The main disadvantage of metals is that they are more complicated to process than wood.

Make and test your own "alloy." Half fill a large paper cup with gravel (stones of similar size and shape). Now fill the cup to the top with water. Half fill another cup of the same size with the same type of gravel. Fill this cup with sand so that it just covers the gravel. Finally, top off the cup with water. Now place both cups in a freezer.

The next day, take your two hardened samples out of the freezer and remove the paper to see their structures. Design your own hardness test for each "alloy." Demonstrate the tests for your teacher.

 Caution: Be sure to wear safety goggles or a face shield while performing your tests.

Mining iron ore underground

Smelting is similar to boiling a liquid; the impurities rise to the top of the molten metal and are skimmed off. Often, metals are mixed with other metals or substances. These new materials are called **alloys**. Steel and brass are examples of alloys. The added metals give the pure metal special properties. For example, in the process of making steel, carbon is added to iron to increase its hardness.

SHAPING

In its molten state, metal can be formed into different shapes. A good example is alloy steel (steel that contains another metal). Molten alloy steel is poured and continually reshaped. The hot steel is rolled and pressed into sheets, beams, rods, and tubes.

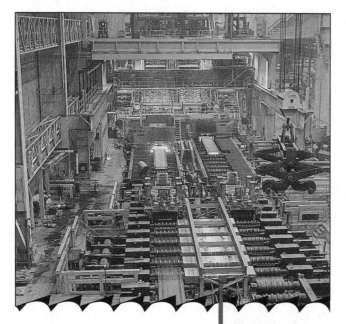

This steel mill shows the complex set-up for casting and shaping steel.

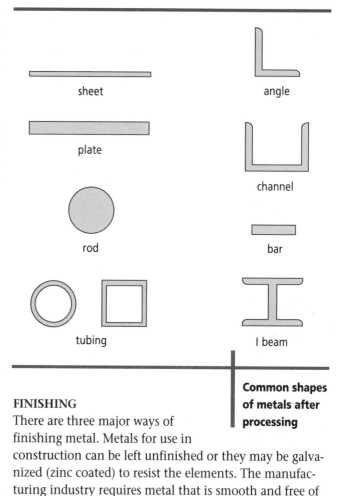

Common shapes of metals after processing

FINISHING

There are three major ways of finishing metal. Metals for use in construction can be left unfinished or they may be galvanized (zinc coated) to resist the elements. The manufacturing industry requires metal that is smooth and free of blemishes. This metal is rolled under pressure to produce a fine, smooth finish suitable for use in the automotive and manufacturing industries. This fine finished metal can be easily painted or chrome plated.

Recognizing and Choosing Metals

Suppose you have decided that metal is the best material for your design requirements. Just as with wood and plastic, there are many different metals. Some come only in certain shapes. You must be able to recognize the metals that are available in most design and technology classrooms.

Metals can be divided into two classes: ferrous and non-ferrous. **Ferrous metals** are metals and alloys that contain iron. Examples are cast iron, steel, and alloy steel. **Non-ferrous metals** contain little or no iron. Examples are aluminum, copper, zinc, silver, platinum, gold, lead, and tin.

Use the chart to help you recognize some common metals. Although your situation may differ, most design and technology classrooms will have alloy steel, aluminum, copper, and lead.

Ferrous Metals

Name	Melting point	Properties	Handling	Common uses
Cast iron	1200–1400°C	weak, brittle, soft under the surface	can be compressed, can be moderately bent, some sparks when grinding, small chips, can be filed	various castings for machines, piston rings
Wrought iron	1600–1700°C	ductile, malleable, slightly soft	bends well in most temperatures, welds well, white curly shavings when drilled, many sparks when grinding, can be filed with more effort	nuts, bolts, wires, tubes, bridges, girders, hardware
Mild steel	1300–1500°C	ductile, less malleable than wrought iron, hard, strong	bends but can fracture with repeated bending, long curled shavings from drilling, grinding causes shower of white sparks, resistant to grinding	nuts, bolts, wires, tubes, bridges, girders, hardware
High carbon steel	1200–1400°C	not ductile, strong, hardness determined by heat treatment	brittle when hardened; resistant to hammer; grinding causes full red sparks; drilling, sawing, filing are difficult	drills, taps, dies, chisels, punches, hammer heads, most hand tools
Stainless steel	1400°C	ductile and resistant to most elements and corrosives	bends well at most temperatures; drilling, sawing, filing can be difficult	kitchen utensils, pots, pans, sinks; used extensively in hospitals, restaurants, and in the chemical and dairy industries

Non-ferrous Metals

Name	Melting point	Colour	Properties	Handling	Common uses
Aluminum	600°C	silver-white	light, malleable, ductile, soft, good electrical conductor	can be bent or formed easily, machines easily, difficult to solder or weld	foil food wrap, car body panels, aircraft engine castings
Brass	8400°C	yellow	soft, ductile	varies according to proportion of zinc and copper, solders easily, polishes well	wire, screws, pipe fittings, castings

(cont'd)

Non-ferrous Metals (cont'd)

Bronze	880°C	yellow-brown	excellent for castings, very corrosion-resistant	machines well, tough	castings, bearing, especially where water is present
Copper	1080°C	orange-pink when polished	malleable, ductile, excellent conductor of heat and electricity	can be bent and formed easily, solders easily	soldering, tubing, wiring, rivets and bits
Lead	327°C	blue-grey with dull shine	very soft and malleable, corrosion–resistant	can be easily bent and formed, solders easily	pipes, roofing, drainage troughs, car batteries
Tin	232°C	silver-white, bright and shiny	extremely ductile, malleable, corrosion–resistant	can be easily bent or formed	used as coating on steel plate, used to form alloys
Zinc	419°C	blue-white	hard, brittle, ductile at low temperature, brittle at high temperature, corrosion–resistant	in sheets, folds easily; can be soldered	used in galvanizing, sheets used in roofing

This house has aluminum siding.

Aluminum is the main material in these computer disks.

Example: Metal for Wind Chimes

In Chapter 4 (see page 50) you learned how to make chimes. Suppose that you are deciding among alloy steel, aluminum, copper, and brass from which to make the chimes. The most important design factors are
- the sound of the metal when it is struck,
- the mass of the metal, since the chimes must be hung, and
- the availability of the metal in tube form, since this shape resonates best.

The only way to discover which metal best fulfills your needs is to test the various kinds. By arranging your information in a rating chart, you may see that aluminum is the best choice for your chimes, with brass coming second. You might design a different chart that gives you a different solution.

Consulting Your Teacher

Once you have decided on the best material, you should consult your teacher to discuss all the factors you considered and how you came to your conclusion. Your teacher may inform you that aluminum and brass are both good choices, but brass tubing costs five times as much as aluminum tubing. For this reason, you choose aluminum for the wind chimes.

Rating metal properties for chimes

	alloy steel	brass	copper	aluminum
Pleasing sound	1	5	1	5
Resonance	3	5	1	5
Lightness	2	2	3	5
Totals	6	12	5	15

Poor = 1   Excellent = 5

The hardness of ceramic tiles makes them ideal for mall flooring.

Ceramics

Ceramics are materials formed from silicon, a non-metallic compound that is abundant in the Earth. Glass is a common ceramic found all around us. One of its main compounds is sand, which contains silicon. It is difficult to believe that the clear pane of glass in a window is made of sand and other compounds. Clay is a form of ceramic that is used to produce objects of art as well as eating, drinking, and cooking utensils.

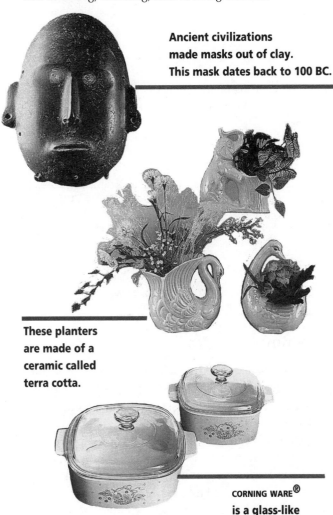

Miniature ceramic chips are the heart of today's high-speed computers.

Ancient civilizations made masks out of clay. This mask dates back to 100 BC.

These planters are made of a ceramic called terra cotta.

CORNING WARE® is a glass-like ceramic cookware.

Ceramics have had practical uses since ancient times. The art of pottery has its origin in early civilizations, when people fashioned utensils and ornaments from clay. Some civilizations built houses of adobe, a brick made of a straw and clay mixture that was dried in the Sun.

Modern technology has changed the traditional application of ceramics. They are no longer limited to pottery, porcelain, bricks, and cement. Although these ceramics remain important materials, new ceramics are used as insulators and in machine tools, computer parts, and automotive parts.

How Ceramics Are Processed

Practically all ceramics are processed by the same method. Natural compounds are crushed, combined, and heated to extremely high temperatures.

Recognizing and Choosing Ceramics

You must consider the requirements of your fabricated project when choosing ceramics. Most ceramics have the advantage of being very inexpensive. The following are the general properties of ceramics:

- very malleable before hardened,
- some types need water for mixing,
- heavy, and
- brittle when dry.

Most of the ceramics you will be using in your design and technology class are easy to recognize. The most common types are glass, clay, plaster, and cement.

GLASS

The different components of **glass**, such as sand, limestone, scrap glass, soda ash, potash, and other minerals, are fused together by the high heat in a kiln. Common uses of glass are windows, mirrors, and jars.

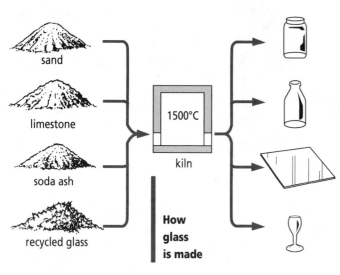

CLAY

A mixture of a very fine type of sand and water makes **clay**. It can be formed into practically any shape. **Pottery** is clay that has been heated to temperatures of 540–590°C in a kiln. The heat changes the clay from a malleable to a rigid state. Pottery is usually finished with a glaze to make it water-resistant and for decorative purposes.

Bricks are made of clay and are an important construction material. They are fired in high heat in a kiln. Bricks are hard, strong, and heat-resistant.

PLASTER

Calcium carbonate (limestone, gypsum, and shale) is the key ingredient in making **plaster**. The calcium carbonate must be dried in a calcining process. This process heats the raw material to the point of partial fusion at a temperature of 1427–1649°C for up to four hours. Calcining is done in a large rotary kiln. The calcined material is ground into a powder (lime or plaster of Paris). The addition of water reverses the process. Plaster is used for dry-

wall in the building industry and moulds and casts for use in hospitals.

CEMENT

Portland **cement** is calcined limestone and red shale. Gravel and sand can be mixed with cement and water to produce concrete. Concrete is used extensively in the building industry.

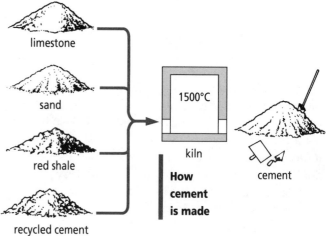

Concrete is a mixture of Portland cement, sand, gravel, and water. The mixture sets to a rigid state and then is cured. Cured concrete is very strong and hard. When it is reinforced with steel, it becomes even stronger.

Ceramics

Name	Appearance
Glass	clear or opaque finished product, hard and brittle
Clay	brown, red, or white powder to which water is added
Plaster	white powder to which water is added
Cement	grey powder to which sand, gravel, and water is added

Recycling Ceramics

The nature of ceramics makes these materials highly suitable for recycling and reusing. Recycling of glass bottles was one of the earliest and most successful environmental programs.

A greater effort is being made to recycle other ceramic products so that they do not fill up landfill sites. Besides bottles, other glass containers can also be recycled, and concrete can be crushed into gravel. Everyone must make a concerted effort to reuse the new ceramics, which will continue to be very popular materials in the future.

Points for Review

- When choosing a material, consider its availability, suitability, cost, and environmental impact.
- All materials possess certain properties, such as hardness, mass, water resistance, heat resistance, fatigue, malleability, elasticity, ductility, brittleness, toughness, and tenacity.
- Standardized testing of different materials will give you the knowledge necessary to choose the correct material for a project.
- Wood is a very popular material since it can be easily shaped, can be cut to various thicknesses and sizes, can usually support a fair mass, can be filed and sanded to accept a pleasing finish, can absorb shock, and can provide insulation.
- Wood, a renewable resource, is divided into the two main categories of softwood and hardwood.
- Plywood, hardboard, and particle board are manufactured forms of wood.
- Plastics are synthetic materials that are manufactured from petroleum substances.
- Plastics are light yet strong and can be flexible or rigid. They can be machine formed, finished, and coloured; they resist rotting and corrosion; and they are effective electrical and thermal insulators.
- Plastics fall into two categories: thermoplastics, which liquefy when heated and solidify when cooled; and thermosets, which solidify when heated and cannot be liquefied again.
- Metal, a non-renewable resource, can withstand heat, conduct electricity, and usually outperform wood.
- Ferrous metals contain iron. Non-ferrous metals do not contain iron.
- Ceramics are formed from silicon, a non-metallic compound that is abundant in the Earth.
- Practically all ceramics are processed by crushing, combining, and heating natural compounds to extremely high temperatures.

Terms to Remember

acrylic
acrylonitrile butadiene
   styrene (ABS)
alloys
annual rings
bark
bricks
cambium
cement
ceramics
clay
concrete
consultation
dressing
epoxy resin
ferrous metals

glass
grain
hardboard
hardwood
heartwood
lumber
materials
metals
moisture content
non-ferrous metals
ores
particle board
physical qualities
plaster
plastics

ply
plywood
polyethylene
polyurethane
polyvinyl chloride (PVC)
pottery
properties
sapwood
smelting
softwood
standardize
thermoplastics
thermosets
veneer
wood

Applying Your Knowledge

1. Imagine that you are designing a new sports product.
   a. Write a description of your sports product (it can be an improvement on an existing product or a completely new product).
   b. List some of the properties mentioned in this chapter that your product will require. Add any properties that have not been mentioned.
2. Form groups of three to do this exercise. Refer to samples of wood supplied by your teacher to:
   a. Identify the surface that shows the grain and the surface that shows the annual rings.
   b. Say whether these samples show knots, warping, checks, wanes, or worm holes.
   c. Discuss which of the samples you would use for a project and which you would hesitate to use. As a group, explain your reasons to your teacher.
3. List five products in your home that are made of plastic.
   a. In a group, report your list. Choose one product from each group member's list and discuss how each product would differ if it were made of wood. What would be its drawbacks?
   b. Assess whether the original choice to use plastic for these products was appropriate.
4. Examine your classroom for different uses of plastics.
   a. List the different needs for plastics and the products that fulfill those needs.
   b. Identify whether the type of plastic used for each product is hard or flexible.
   c. Which of the plastics do you think may be thermosets?
5. In a class discussion, debate the use of recyclable versus non-recyclable plastics. What qualities of non-recyclable plastics make them difficult to be replaced by recyclable ones? You may need to do some research into both types.
6. On your way to school, make notes on the different uses of metals you see.
   a. List four items that are made of metal.
   b. Write down the characteristics of each item.
   c. Note the characteristics that are common to all four items.
   d. Discuss with a partner the possible reasons why each item was made of metal instead of wood or plastic.
7. The automotive industry is using more and more plastic. In groups of three, list the benefits and drawbacks of this increased use of plastic. Present your list in a class discussion.
8. Both glass and clay are ceramic materials. Write a brief paragraph on why you think these two materials are put in the same category.
9. In groups of three, have a brainstorming session to help solve the following problem: You need to choose a material to make a pot that will hold other materials at high temperatures.

Chapter 9

FABRICATION PROCESSES

What You Will Discover

After completing this chapter, you should be able to:

- Understand the importance of fabrication in the design process.
- Identify the best fabrication processes for your design.
- Apply your knowledge of materials to your choice of the best fabrication process.
- Understand the need to measure materials correctly.

*F*abrication can be defined as the process of building, constructing, or manufacturing. It is a crucial stage of the design process, where materials and parts are put together according to plan. A good knowledge of the fabrication methods for the materials you will use will certainly make your task easier.

In this chapter, you will learn about the different types of fabrication processes for woods, plastics, metals, and ceramics. You will also learn about the tools and machines available in most design and technology classrooms that will allow you to perform these processes.

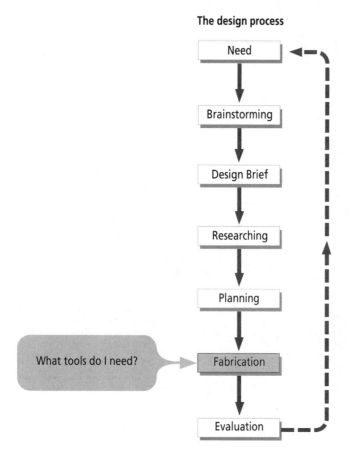

The design process

Need

Brainstorming

Design Brief

Researching

Planning

What tools do I need? → Fabrication

Evaluation

Choosing a Fabrication Process

*I*n Chapter 2, you learned that choosing the right fabrication process is part of planning. You must first choose your material — only by making this decision can you begin to decide the basic processes you might use.

Recall from Chapter 8 that the four basic materials are woods, plastics, metals, and ceramics. These are the materials to which you will be applying at least one of the basic fabrication processes: measuring and marking, cutting, bending, pouring, joining, and finishing. With

some important exceptions, all of these processes can be applied to the four types of materials.

The chart at the top of page 149 shows the particular ways in which each material can be processed (measuring and marking are discussed in a separate section because they are very important). For example, if you want to know the ways that wood can be bent, look across the "Woods" row and down the "Bending" column. The box tells you that wood can be bent by soaking it in water or by steaming it.

You will notice that in some cases different materials can be processed in the same way. For example, woods and plastics can be sawed and drilled. In other cases, a process is exclusive to particular types of materials. For example, woods are the only materials that cannot be poured.

Using Available Tools

*W*hether you can perform one of the basic fabrication processes on a material depends on the available technology — the tools needed to fabricate the project.

Successful designers are economical in their use of resources. The resources include not only materials, but also tools. By minimizing the amount of resources used and ensuring that the fabricated project meets your needs, you make the most of all the resources that go into making a project. This means that there will be little wastage, which is good for the environment. It also means that the project will be more cost-effective.

In the classroom, you will usually have a wide selection of tools and machines to use in fabrication. The bottom chart on page 149 lists the six basic fabrication processes and the technology they use. You should check that all the tools you need are available. Talk to your teacher if a particular tool you need is not available. Your teacher will have already assessed the tools that need to be replaced or purchased; only rarely can you expect your teacher to purchase large quantities of tools during the school year.

Consulting Your Teacher

*W*hen you consult your teacher about your material choice, you should also discuss the possible fabrication processes. In general, you should discuss
- the material you have chosen, and why (see Chapter 8),
- the planned ways of putting the parts of your project together, using sketches or models, and
- the bill of materials, including all sizes of materials needed.

Fabrication in the Design Process

Material	Cutting	Bending	Pouring	Joining	Finishing
Woods	• saw • drill • chisel • plane • file • rout	• soak in water • steam	• cannot be poured	• mechanical • bonding • cut and fit	• sand • coat
Plastics	• saw • drill • mill	• strip heat • vacuum form • thermoform	• cast • mould	• mechanical • bonding	• sand • file • buff • polish • coat
Metals	• saw • drill • mill • lathe • shear • punch	• hammer • forge • press in a jig	• cast	• mechanical • bonding	• file • coat
Ceramics	• saw • drill • fracture	• hand mould • shape	• mix in raw state	• mechanical • bonding	• coat

Common Classroom Technology for the Basic Fabrication Processes

Measuring and Marking	Cutting	Bending
• scribers • rulers • tape measures • marking gauge and dividers • squares • levels • callipers • micrometers	• clamps • files • scissors, snips, and shears • chisels • planes • saws • drills • sanders • routers • mills	• hammers • moulds • strip heaters • jigs • vises and clamps

Pouring	Joining		Finishing
• moulds	**Mechanical Joints** • saws • glues • vises and clamps **Bonding** • glues • clamps • solders • soldering iron • welding tools	**Mechanical Fasteners** • nails • hammers • screws • screwdrivers • nuts and bolts • rivets • riveting pliers • moving joinery	• sandpaper • sanders • files • coatings • brushes

Career Profile

Swanny Graham

"In today's economy, the more versatile you are, the more solutions you can apply to any given problem, whether it be finding new clients or designing new products," says Swanny Graham, an industrial designer and entrepreneur. "Only a flexible mind allows you to see solutions from many angles."

After completing university, Graham began working with a large company, then moved to a small company. The two experiences gave her a "more balanced view of what design and manufacturing is all about."

In the large company, she gained experience working with a large team of professionals. The company manufactured electronic "reading and sorting equipment. For each part of the equipment ... several designers, technicians, engineers, and clerical people would be involved in the long design process."

The small manufacturing company had a plastics shop and a woodworking shop. "We designed and manufactured display systems for stores. I would be involved with the customer, the design of the product, the manufacturing, and the distribution," says Graham. "I had to come up with quick creative solutions, often in front of the customer — but always with the practical in mind, because I would also be responsible for the manufacturing. I had to learn to make fast decisions, to learn quickly from my mistakes.

"I recall having to design a metal display unit that would be in stores all across Canada. It was a huge order for the company. But we only had a few days to design it properly and then to start production. The project worked out fine because of everyone's incredible efforts on the team."

Now Graham is using her varied experience as an entrepreneur. "Recently, an associate and I have set up our own design company. We had worked together before and knew each other's strengths and weaknesses. We also knew that we worked well together when it came to creating new ideas and producing finished results within a limited time. Presently we are working on several proposals for different clients. One is a metal and plastic product for the medical field, and another is a unit for displaying retail products.

"For several years I have also been developing a few products on my own. They are at various stages of development. They range from an angle-measuring device, a lightweight scale, a geometry game for the visually impaired, and a product to encourage recycling of food jars."

For anyone thinking of entering the field of industrial design, Graham emphasizes the necessity of lifelong education. "I have continued to take courses every year, either full-time or part-time. Each course has been either creative or technical. Integrating the creative approach with the technical approach has given me the skills to discover a need for a product and to apply the processes of design to that potential product. I can also come up with several creative options that are also realistic because of my knowledge of materials and manufacturing. Education should be ongoing, regardless of your age or position in your profession. Shortly I will be taking part-time engineering courses at the University of Western Ontario."

Graham also stays alert to today's environmental issues. "The overproduction of products over the last several years has increased the public's concern and awareness about recycling and the environment. This presents many challenges for creative thinkers to come up with products that provide both value and satisfaction to the consumer — and at the same time address environmental concerns."

QUESTIONS AND EXERCISES

1. Why is versatility important when designing?
2. When designing a metal display unit, why did Swanny Graham have to quickly come up with a final design?
3. What are the products that Swanny Graham creates?
4. How has integrating creative and technical approaches assisted Swanny Graham when designing products?
5. Sketch five possible products that you would design if you worked with Swanny Graham.

Refer to the sketches, drawings, model, and notes you made during the brainstorming and researching steps. Indicate which of the six basic fabrication processes — measuring and marking, cutting, bending, pouring, joining, finishing — you will be using. If you will be using more than one, you will have to decide the order of the processes when consulting your teacher. Refer to the classroom technology chart at the bottom of page 149 to decide what you will need to perform the necessary processes.

When you reach an agreement with your teacher on the materials and fabrication processes, you will be ready to begin fabrication.

Testing Your Fabrication Process

Once you begin to fabricate a project, it is sometimes difficult to change a procedure. Changing a step can add cost and time to your entire design process. For this reason, material tests and models are helpful in pointing out any unforeseen problems, and they can speed up fabrication.

Finalizing Fabrication Steps

When you are satisfied that you will be using the best material and the best fabrication processes in the best order, be sure to list the sequence of steps on a fabrication chart.

You may have already filled out such a chart and changed it several times. Be sure it is up to date based on all your final decisions. Keep the chart handy for quick reference to make sure that you are doing the right thing at the right time.

The same applies to your drawings of the design. Keep copies with you always to refer to whenever you want to check your measurements and overall design.

Basic Fabrication and Tools

This section will give you a good idea of the *general* types of technology that allow you to carry out fabrication processes. All the variations of these basic processes cannot be explained, since there are literally hundreds of tools and machines available for fabrication.

Measuring and Marking

You must establish a reference line or point from which to measure your material. By **measuring** and **marking**

accurately you will ensure that the following will likely occur:
- your design will be the size you planned,
- you will not waste material, and
- your cutting, bending, and joining will be accurate.

You will need to measure and mark at least once during the design process. It is a good practice to take measurements three times in order to minimize errors. When reading a measurement, make sure your eye is directly above the number on the ruler or measuring tape.

Measuring and Marking Tools

The following are common measuring and marking tools found in the average design and technology classroom.

PENCILS
Use a sharp pencil to mark your measurements on a material. Remember that you will be erasing these pencil marks during the finishing stage, so do not press too hard.

SCRIBERS
The **scriber** works like a pencil, but it does not make a coloured mark. It is used to scribe, or scratch a line into, materials — typically plastics, metals, and ceramics. The scriber is made from hardened steel to resist wear.

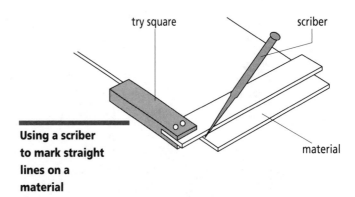

Using a scriber to mark straight lines on a material

RULERS AND MEASURING TAPES
Rulers can be made of wood, plastic, or steel and can have unit markings etched on one or both sides. A 30.5 cm ruler can be used to measure most work. Where longer measurements are involved, you can use a **measuring tape**. This is a longer, flexible type of ruler.

MARKING GAUGES AND DIVIDERS
These are tools that scribe a fine line into a material such as wood, plastic, or metal. A **marking gauge** scribes lines parallel to an edge. The movable head can be adjusted to different widths and can mark out the same width on several pieces of material quickly and accurately.

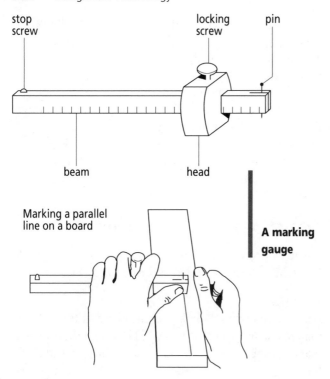

stop screw

locking screw

pin

beam

head

A marking gauge

Marking a parallel line on a board

A set of **dividers** works like a compass. The two points are used to obtain a measurement and scribe it into a material. Dividers are also used to mark off equal spaces.

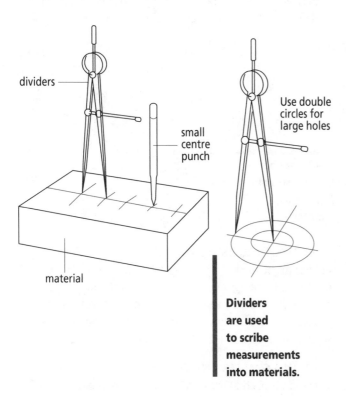

dividers

small centre punch

Use double circles for large holes

material

Dividers are used to scribe measurements into materials.

CENTRE PUNCHES

The **centre punch** is very useful when drilling holes. It is used to mark the centre of a hole before drilling.

TRY SQUARES

The **try square** is used to test the squareness of a material. It is also used to draw cutting lines that are perpendicular (90°) to an edge.

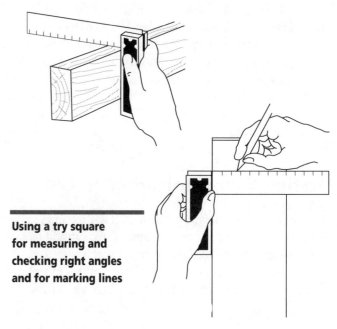

Using a try square for measuring and checking right angles and for marking lines

COMBINATION SQUARES

A **combination square** has a movable head that slides on a grooved ruler and can be tightened at different points. It is used for squaring the end of a material, measuring several points on a material, and drawing 45° and 90° angles.

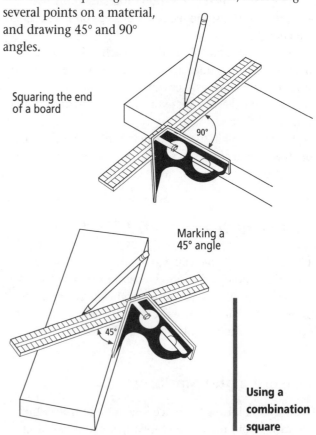

Squaring the end of a board

90°

Marking a 45° angle

45°

Using a combination square

LEVELS

The **level** is used to determine whether a surface is perfectly horizontal or vertical. The level has horizontal and vertical tubes that contain a liquid with an air bubble inside. When the bubble lies directly between the lines on the tube, it shows that the surface is level.

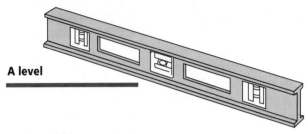

A level

CALLIPERS

There are two types of **callipers**. Outside callipers are used to measure the outside dimensions of round surfaces. Inside callipers measure the inside dimensions of a hole. Callipers are used extensively on the lathe to measure and to transfer measurements from a drawing to a material.

 Caution: If you are using a lathe, be sure to turn it off before using these and other measuring tools.

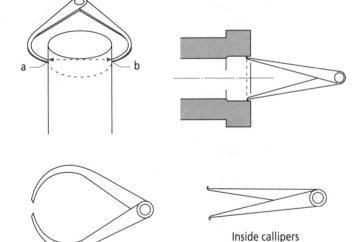

Outside callipers

Inside callipers

ADVANCED CALLIPERS

Both the **micrometer** and the **vernier callipers** combine callipers and a ruler. These measuring tools allow you to measure with great precision (for example, the thickness of a hair).

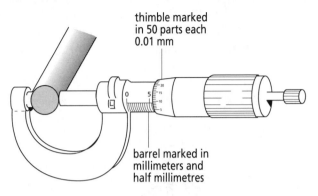

thimble marked in 50 parts each 0.01 mm

barrel marked in millimeters and half millimetres

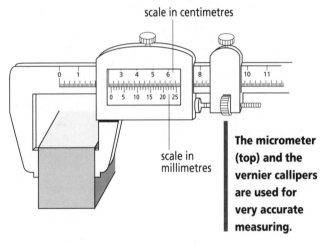

scale in centimetres

scale in millimetres

The micrometer (top) and the vernier callipers are used for very accurate measuring.

Cutting

Cutting is piercing through or shearing off part of a material to achieve an overall shape. You will use a variety of tools to cut materials.

 Caution: All cutting is best done with sharp tools. Be sure to follow all the safety precautions explained by your teacher before using any cutting tool.

See for Yourself

Assess your ability to take accurate measurements. With a partner, select a piece of wood supplied by your teacher. Using a ruler or measuring tape, find the length of the wood. Write down the measurement in your notebook. Do this two more times so that you have three measurements recorded in your notebook. Now have your partner measure the length three times. Compare your results. Were all the measurements the same? What is the agreed length? What do the results tell you about the accuracy of taking only one measurement?

Cutting Tools

CLAMPS

Clamps do not cut but are often used to assist when cutting a material. They hold a material still against a surface such as a workbench. This keeps the material from slipping as you are cutting it. You should put pieces of wood between the jaws of the clamp and your material to avoid damaging the material. Consult your teacher before you cut to find out whether you should use a clamp and which type of clamp to use.

The **bench vise** is a clamp that is mounted on a workbench.

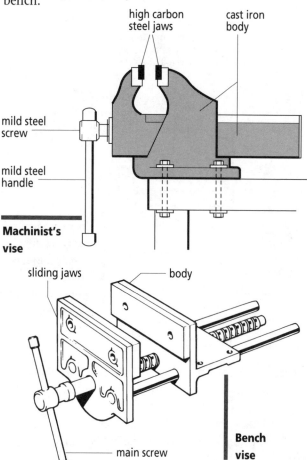

high carbon
steel jaws

cast iron
body

mild steel
screw

mild steel
handle

**Machinist's
vise**

sliding jaws

body

main screw

**Bench
vise**

The **C-clamp** is shaped like a C. The frame is made from forged steel and is very strong. This clamp is very versatile and can be used to clamp work of different thicknesses and widths.

C-clamp

The **bar clamp** is used to clamp large-surface and frame work. It has two adjustments: coarse and fine. The coarse adjustment is made by sliding the second jaw forward or back. The fine adjustment is made with the handscrew only after the coarse adjustment has set the clamp snug to the work.

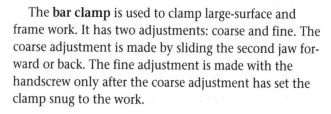

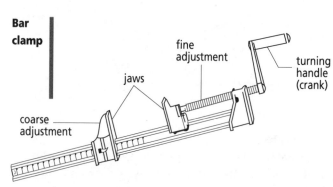

**Bar
clamp**

fine
adjustment

turning
handle
(crank)

jaws

coarse
adjustment

The **handscrew clamp** has jaws that can clamp irregular angles. To adjust this clamp, grip a handle in each hand and turn them. Open the clamp until it barely fits over the work. Close it at the heel and slip the jaws over the work. Open the clamp again at the heel and the jaws will close tightly on your work. Install handscrew clamps on both ends of your work to provide equal clamping pressure.

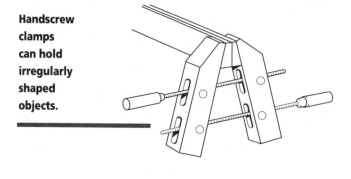

**Handscrew
clamps
can hold
irregularly
shaped
objects.**

FILES

Files are extremely versatile tools that can be used on woods, plastics, and metals to remove excess material or to finish a surface. They can also remove saw marks and burn marks in materials.

Files are made of high-carbon steel, which is hardened and tempered. The most common shapes are flat, half-round, round, triangular, and square. Each shape is used to file surfaces that have the same shape. The markings, or cut, of a file also determine its use. Files should not be used when they are dull.

With the **draw-filing** technique, hold the file diagonally and draw it along the work. This produces smooth, flat surfaces. With the **cross-filing** technique, hold the file horizontally and draw it along the length of the material. This method is preferred when material removal is more important than finish.

 Caution: Always use a file with a handle; otherwise the tang may injure you.

Shapes and parts of a file

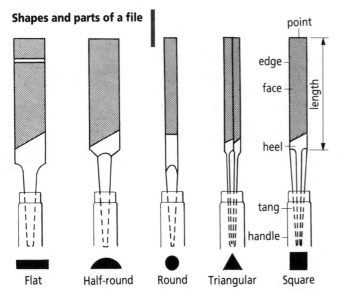

Flat   Half-round   Round   Triangular   Square

Files must be used carefully as they are brittle and can be easily damaged. They should also be cleaned after use.

A **rasp** is a rough file that can cut curves and straight lines as well as sculpt wood. It is used for rough work on wood, before you use a file for finishing. It is easy to use, and a lot of work can be accomplished in a short period of time.

SCISSORS, SNIPS, AND SHEARS

Scissors are common tools that can be used to cut paper and cardboard when making models.

Snips are heavy-duty scissors used for cutting thicker and harder materials, such as sheet metals and some plastics. There are different snips for cutting straight lines and right and left curves. After you scribe cutting lines into a material, cut on the waste side, leaving the line on the work until final finishing.

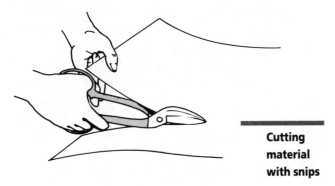

Cutting material with snips

A **beverly shear** is a large, table-mounted cutter (snips) that is used for cutting heavier sheet metals. This tool has a very long handle, which provides the mechanical advantage of a lever. Beverly shears are restricted to cutting straight lines.

Wire cutters are small snips used to cut wire, tin, and other soft metals.

CHISELS

Different kinds of **chisels** are used for accurate cutting and shaping, usually of woods and metals and sometimes plastics. They may also be used to carve a straight line into, or score, ceramic tile as a first step in cutting. The glass cutter is a special type of rotary chisel.

Carving chisels are used to sculpt wood. You should always carve with the grain and away from your body. Different shapes of chisels give different cuts; use the appropriate chisel to attain the desired effect.

There are also special chisels that are made of hardened steel for cutting metal. A ball-peen hammer is used to drive these chisels.

 Caution: When using a chisel, make sure your work is securely clamped.

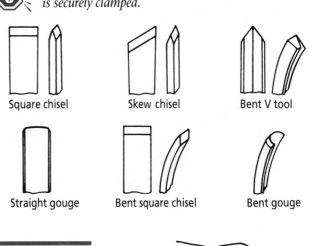

Square chisel   Skew chisel   Bent V tool

Straight gouge   Bent square chisel   Bent gouge

Using a chisel

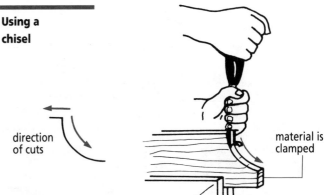

direction of cuts

material is clamped

PLANES

A **plane** is used to square and shape wood. It is really a chisel mounted in a rigid handle. A good plane is very helpful for smoothing material. This tool is most often

**See for Yourself**

A good way to learn how to make a sheet-metal box is by making a paper model. You will need paper, tape, and scissors. Before you make your own box, examine how a cereal box is constructed. Carefully detach the glued parts and lay the open box flat. Count the different shapes that you see. Can you see a pattern where the same shapes occur? Using these observations, design and fabricate your own paper box.

restricted to straight cuts; the larger the plane, the straighter the cuts will be.

Parts of a plane

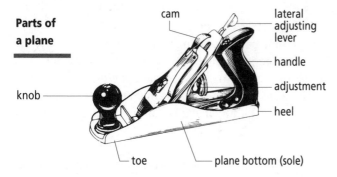

Secure your work to a bench with a vise or clamp. Plane with the grain of the wood, otherwise you may tear it. A very sharp blade and the correct adjustments will make the task of planing wood very easy.

Using a plane

SAWS

A variety of **saws** are used to cut different materials. **Saw pitch** describes the number of teeth/25 mm, which indicates the coarseness or fineness of a saw. Pitch is important when cutting different thicknesses and types of materials. Hard and thin materials are best cut with a large-pitch saw (a greater number of teeth); soft and thick materials are best cut with a small-pitch saw (a smaller number of teeth).

This saw has a pitch of 7 teeth / 25 mm

(7 teeth/inch)

In general, hand saws are used to cut materials that are fairly soft. When using a saw, secure your work firmly and cut on the waste side. Leave the cutting line on the work so that you can finish to this line using a file or a plane, depending on your material.

The **crosscut saw** and the **rip saw** are used to cut wood. The crosscut saw is used to cut across the grain of the wood, and the rip saw is used to cut with the grain.

Parts of a crosscut saw

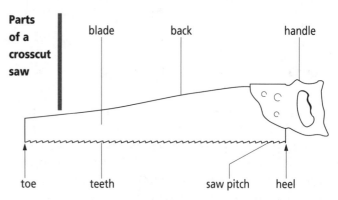

To use the crosscut and rip saws:
1. Support the lumber to be cut, using a sawhorse or a vise.
2. One hand holds the saw. The other hand steadies the material.
3. The saw teeth should be on an angle to the surface being cut. A 45° angle is needed for crosscutting and a 60° angle is needed for ripping.
4. Start the saw with short drawing strokes.
5. As cutting progresses, try to use the full length of the saw, except for the toe and the heel.
6. To prevent splitting the wood as the cut reaches the end, hold the free end of the lumber and ease up on the pressure being exerted on the saw strokes.

The **backsaw** is used for making accurate cuts in wood across end grain and for joinery work. It has a thin blade supported by a straight steel back. A mitre box is used to guide the backsaw when making joinery and angle cuts. The **coping saw** is a specialty saw for cutting irregular shapes and curves in wood. The **hacksaw** is used to cut plastic or metal, depending on the blade used.

Caution: *When using a hacksaw to cut metal, be very careful not to cut yourself on the edge of the metal — it can be very sharp.*

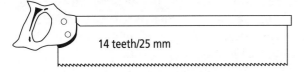

14 teeth/25 mm

Backsaw: used for
straight cuts in wood

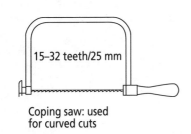

15–32 teeth/25 mm

Coping saw: used
for curved cuts

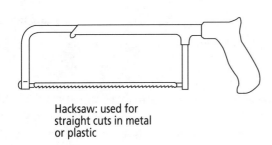

Hacksaw: used for
straight cuts in metal
or plastic

ELECTRICALLY POWERED SAWS

Electrically powered saws have the advantage of being able to make the cutting edge move very fast. The most common type is the **sabre saw**. This saw has variable speeds and can cut woods, plastics, and metals with its wide selection of replaceable blades.

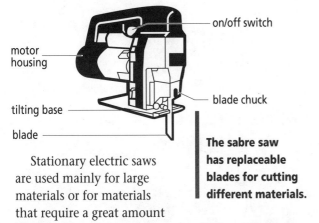

on/off switch

motor
housing

tilting base

blade

blade chuck

The sabre saw has replaceable blades for cutting different materials.

Stationary electric saws are used mainly for large materials or for materials that require a great amount of effort to cut with hand or portable saws. The common types are the scroll saw, the band saw, and the table saw.

The **scroll saw** cuts woods, plastics, and metals using different speeds and types of blade. This saw can cut intricate shapes, puzzles, and ornaments easily. The scroll saw can also accommodate a file accessory.

The blade of the scroll saw is held by an upper and a lower chuck. The blade cuts on the downstroke; the upstroke is simply a recoil action — it does not cut. Blades come in different sizes, widths, and saw pitches. The width of the blade determines the size of curved cuts; narrower blades can cut finer curves. Blades are installed with the teeth pointing downward. You must use the pressure fingers and hold your work on either side of the cutting line. Sometimes the blade will stick and cause your work to bounce up and down. The reasons could be that you are trying to cut too sharp a curve, you are trying to cut too fast, you have a dull blade, or the pressure fingers are not resting on your work. Scroll-saw blades must be changed frequently. The up and down movement of the blade only exposes about 5 cm of the teeth to the material. These teeth become dull with frequent use.

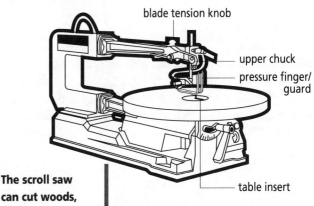

blade tension knob

upper chuck

pressure finger/
guard

table insert

The scroll saw can cut woods, plastics, and metals, depending on the cutting speed and the blade used.

The **band saw** is used to cut mainly woods and plastics. Metals are cut on a band saw equipped with a metal cutting blade. The material you cut must be of a minimum size (check with your teacher) and must be placed flat on the table. The band saw blade is a continuous band that rotates around an upper wheel and a lower wheel. The blade continuously cuts in a downward motion; this holds the work to the table. A light pushing action will cause the blade to cut into your material. Your hands should always be on either side of the cutting line. A push stick should be used when there is danger of your hand coming close to the blade. A dull blade makes cutting difficult and can burn wood; have your teacher check the blade when this happens. The band saw can cut reasonably straight lines as well as curved lines. A fine, narrow blade is needed for curved cuts.

Caution: *This saw must be used with extreme care.*

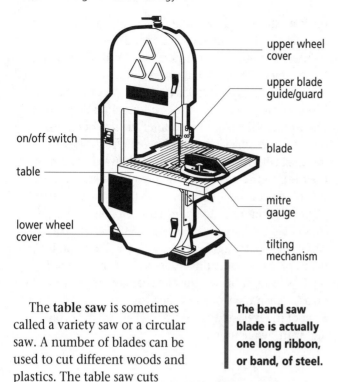

upper wheel cover

upper blade guide/guard

on/off switch

blade

table

mitre gauge

lower wheel cover

tilting mechanism

The **table saw** is sometimes called a variety saw or a circular saw. A number of blades can be used to cut different woods and plastics. The table saw cuts wood in two ways. Crosscutting requires cutting across the grain of wood. Use the mitre gauge only. Both hands must be on the mitre gauge, holding the wood to the flat face. Move the mitre gauge and wood past the blade to make a cut. You must follow minimum lengths and maximum widths of material for this operation (check with your teacher).

Making long cuts, or "ripping" wood, requires the use of the fence and a push stick. The wood has to be of a minimum length and maximum width (check with your teacher). When ripping, use the long reach of the push stick to push the wood. The push stick acts as an extension of your hand.

The band saw blade is actually one long ribbon, or band, of steel.

 Caution: The mitre gauge and the fence are two separate and distinct parts of the table saw. Use them separately (never together — your material could bind; and then bounce back and injure you).

A table saw

DRILLS

Drills are typically used to make holes in woods, plastics, metals, and ceramics. The fast rotation of the **drill bit** causes the cutting action. Various bits drill different sizes of holes and are made for different purposes. Drilling ceramics requires a carbide-tipped drill bit. Drill bits have to be tightened securely in the chuck.

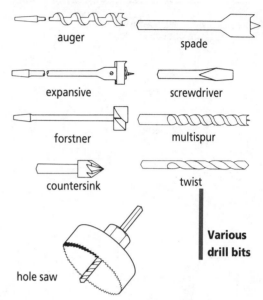

auger

spade

expansive

screwdriver

forstner

multispur

countersink

twist

hole saw

Various drill bits

In general, the harder the material to be drilled, the slower the bit should turn. Make sure that all materials to be drilled are securely clamped. Also, you should have a piece of scrap material behind the material being drilled.

The **hand drill** is a slow-moving manual drill best used on materials that are not too hard. The **hand brace** is another manual drill; it is used to drill large holes in wood.

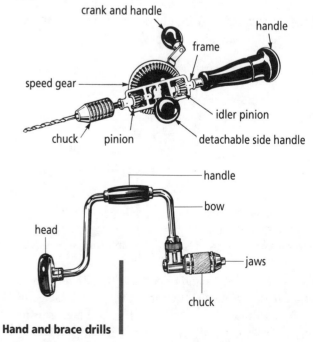

crank and handle

handle

frame

speed gear

chuck

pinion

idler pinion

detachable side handle

handle

bow

head

jaws

chuck

Hand and brace drills

ELECTRICALLY POWERED DRILLS

Electrically powered drills have the main advantage of greater speed. Drill bits can get hot if they are used at too high a speed, are dull, or have not been used with coolant cutting fluid when drilling metal. Excessive heat can damage drill bits.

The **portable electric drill** is inexpensive and very versatile. This is a good, all-purpose drill for woods, plastics, metals, and ceramics, since it can have variable speeds. It can be used to drill holes, install and remove screws, sand, file, polish, and strip finishes. The portable electric drill is powered by electricity using a cord or a battery pack.

Portable electric drill

The **drill press** is used for drilling holes more accurately in woods, plastics, metals, and ceramics. The entire working mechanism is in the head of the machine. Small-diameter drill bits require high speeds and large-diameter drill bits require low speeds. Before drilling materials other than wood, check with your teacher for the correct drill speed and drill bit.

Caution: Material on the drill press must be securely held with a clamp. Should the material slip, it will spin like an airplane propeller and may cause injury.

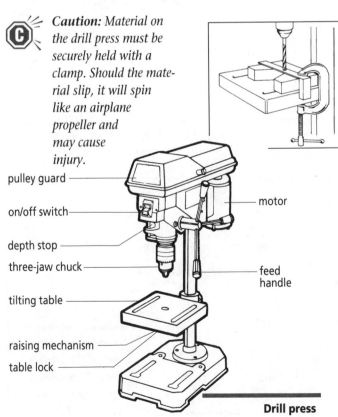

pulley guard

on/off switch

depth stop

three-jaw chuck

tilting table

raising mechanism

table lock

motor

feed handle

Drill press

ROTARY SANDERS

Electric rotary sanders use rotating action to move sandpaper quickly. To cut excess material, the sander is fitted with sandpaper that has a coarse grit. For finishing work, finer sandpaper is used. Sanders are usually used on woods and plastics, and sometimes on metals. Some models have a built-in dust collector bag. The two types of rotary sanders are the portable **belt sander** and the stationary **belt/disc sander**.

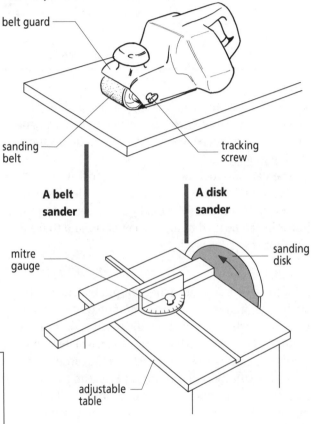

belt guard

sanding belt

tracking screw

A belt sander

A disk sander

mitre gauge

sanding disk

adjustable table

ROUTERS

Electric **routers** are very similar to portable drills. The router is a high-speed motor containing a chuck that can hold a variety of cutting bits to shape wood. By adjusting the depth ring, the desired depth of cut can be achieved. The router is designed to be held with both hands. The base rests on the material, and as the router bit spins over the work, it cuts a shape. The bits can be used to carve a name or shape an edge. A more advanced type of router is the **CNC router**, which is controlled by a computer (CNC stands for computer numerical control).

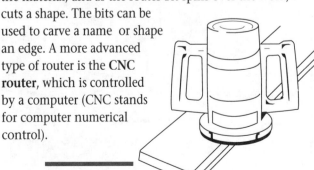

A router

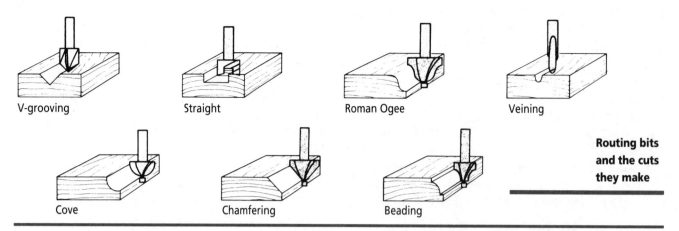

V-grooving   Straight   Roman Ogee   Veining

Cove   Chamfering   Beading

Routing bits and the cuts they make

MILLS

The **mill** is used to shape woods, plastics, and metals in much the same way as very sophisticated woodcarving tools or routers shape wood. The mill can hold various cutting tools.

The standard tool for most school applications is the vertical mill. This relies on a table with T-slots, which allows you to mount the work by a number of methods. The mill can be used to drill, to cut gears, and to make parts for your projects.

LATHES

All **lathes** use fast rotary action to spin materials as they are being cut. A material can be mounted on a lathe at both ends, which is called spindle turning, or it can be mounted at one end, which is called face-plate turning. A chisel-like cutting tool is pressed into the turning material — the cutting action is much like peeling a potato. In almost all cases, lathes are used to shape cylindrical objects, such as bowls, candlesticks, chair legs, metal posts, wheels, and axles.

There are two general types of lathe, the **wood lathe** and the **metal lathe**. The main difference between the two is that the metal lathe has the cutting tool mounted on a carriage. On the wood lathe, the operator must hold the cutting tool. The metal lathe can be used to turn metals, plastics, and woods.

Parts of a metal lathe

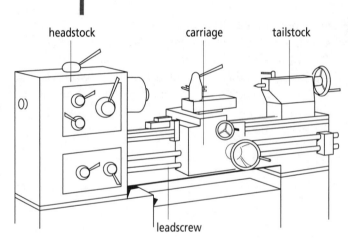

headstock   carriage   tailstock

leadscrew

Different types of cutting tools make differently shaped grooves. These tool bits (chisels) sit on a tool post that is guided slowly into the turning material.

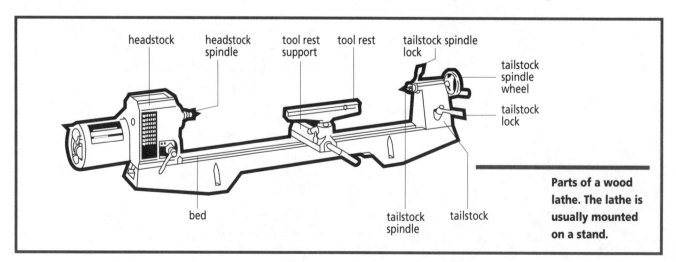

headstock   headstock spindle   tool rest support   tool rest   tailstock spindle lock

tailstock spindle wheel

tailstock lock

bed   tailstock spindle   tailstock

Parts of a wood lathe. The lathe is usually mounted on a stand.

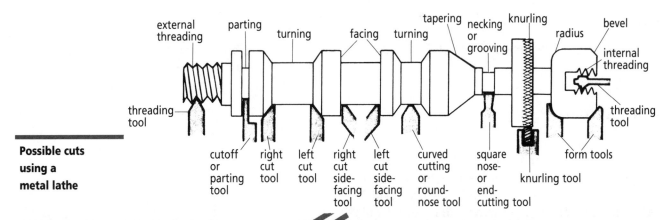

Possible cuts using a metal lathe

Labels: external threading, parting, turning, facing, turning, tapering, necking or grooving, knurling, radius, bevel, internal threading

threading tool, cutoff or parting tool, right cut tool, left cut tool, right cut side-facing tool, left cut side-facing tool, curved cutting or round-nose tool, square nose- or end-cutting tool, knurling tool, form tools, threading tool

Career Profile

Jacques Giard

"Never once did I imagine that my hobbies could become my profession," says designer and educator Jacques Giard. "As a teenager, I spent a great deal of time dreaming, painting, and making scale models, but who would have believed that these same activities would eventually serve the fundamentals of my career as an industrial designer?"

Like many teenagers, Giard became keenly aware in high school that he had to make a career choice. In other words, he had to answer the daunting question, "What do you want to be when you grow up?"

"For me there was no easy answer. My academic strength was science; my personal love was art," says Giard. "And in no way could I see a marriage of these two fields of interest."

Industrial design was largely an unknown profession to Giard as well as to his peers and teachers. "When the time came to make the inevitable career decision, my head took me toward engineering and my heart toward fine arts. It seemed that it was impossible to have both," he says. Yet neither engineering nor fine arts were where Giard settled. Accepted into an engineering program, Giard turned the offer down. Convinced that he wanted to study fine arts, he applied to a program and was rejected. "My only alternative was design, or as it was known then, applied arts," says Giard. "The rest, as the saying goes, is history."

After university, Giard worked as an industrial designer in Montreal. He helped design furniture for homes and offices, as well as wooden toys. Later he worked for a provincial research centre, where he designed equipment for surgery, such as operating tables. Then, as an employee of another design firm, he was part of a team that designed many of the accessories for the Laser sailboat. Giard is now devoting him-

self full-time to education. He is the program chairman at the School for Industrial Design at Carleton University, but he has not abandoned his love of design.

"I have continued to exercise those things that I loved as a teenager — drawing, painting, model making, and more — for over twenty years. And I cannot think of a more exhilarating or satisfying endeavour for any creative person who possesses a streak of practicality. If you must spend a great deal of time in a career, let it be something that you enjoy thoroughly, something that makes you look forward to Mondays."

QUESTIONS AND EXERCISES

1. Which two professions does Jacques Giard believe are blended in industrial design?
2. List the different industrial design positions that Jacques Giard has held.
3. Why is industrial design such a rewarding field to work in?

Bending

Bending is changing the original shape of a material to achieve a new form.

Bending Tools

HAMMERS

The **hammer** is a common hand tool that is used to bend materials, mostly metals. One such way is in riveting: the protruding metal of a rivet is bent using the flat end of the **ball-peen hammer**. A flat head is formed by carefully striking the rivet with the ball end of the hammer. When a number of rivets are used, try to keep the head shapes uniform. (For more on riveting, see page 167.)

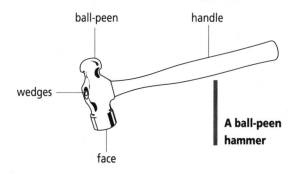

ball-peen   handle

wedges

face

A ball-peen hammer

A **rubber mallet** is a special type of hammer that can be used to bend sheet metal. Since the head is made of rubber, there is little chance of denting the metal.

MOULDS

The process of **thermoforming** involves heating a plastic sheet until it softens. The hot, flexible plastic is forced against the contours of a **mould** by mechanical means (using plugs, solid moulds, and tools) or by pulling it with a vacuum. Boat hulls and plastic sinks are examples of products made this way.

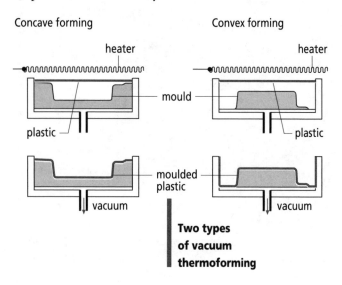

Concave forming

Convex forming

heater

heater

mould

plastic

plastic

moulded plastic

vacuum

vacuum

Two types of vacuum thermoforming

STRIP HEATERS

Sheet plastics can be bent using a **strip heater**. A line is marked on the plastic to show where the bend will be made. The plastic is then placed over the strip heater with the bending line directly over the heating element.

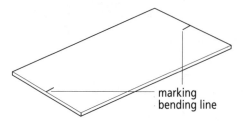

marking bending line

It is important to heat the plastic evenly before bending. Do not overheat it, as this will make the plastic brittle and cause it to crack. Overheating will also cause the plastic to bubble. Once the plastic is soft, it can be bent to the desired angle.

Caution: Remember to wear heat-resistant gloves when using this process.

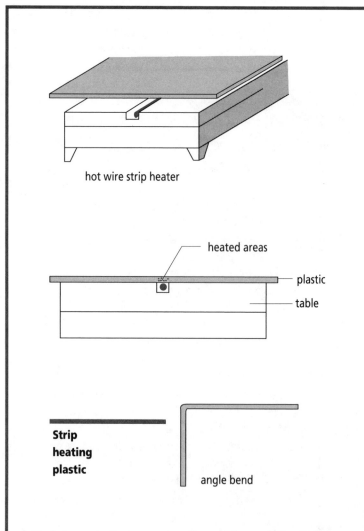

hot wire strip heater

heated areas

plastic

table

Strip heating plastic

angle bend

JIGS

Woods, plastics, and metals can be shaped with **jigs**. Once a piece of wood is steamed or soaked in water, it becomes very pliable. When clamped to a jig and allowed to dry, the wood takes the shape of the jig. The curved backs of chairs are formed this way.

 Caution: Steam can cause burns. Steaming must be done with the supervision of your teacher.

Plastics and thick metals can also be heated and shaped in a jig. Thin metals can be formed by pressure in a jig.

The **box and pan brake** is a table-mounted jig with a lever that is used to bend sheet metal. The metal is inserted, and when you lift the lever, the metal bends.

Pouring

Pouring is using material in a liquid state to make a final solid shape.

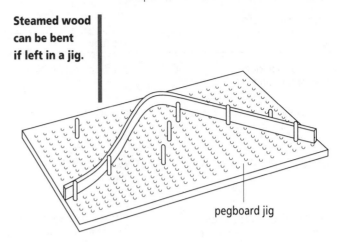

Steamed wood can be bent if left in a jig.

pegboard jig

Pouring Tools

MOULDS

Metals with a low melting point, such as aluminum, can be heated to a liquid state and poured into a **moulding box**. Simple metal shapes can be cast by shaping the sand that fills the box. The liquid metal is poured into the sand cavity through the sprue hole. When the metal fills the cavity it will appear in both the sprue hole and the riser hole. On cooling, the metal can be removed.

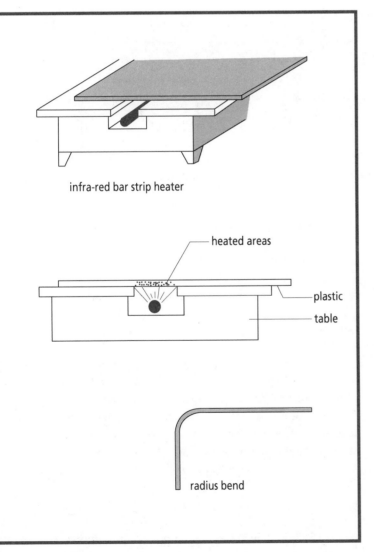

infra-red bar strip heater

heated areas

plastic

table

radius bend

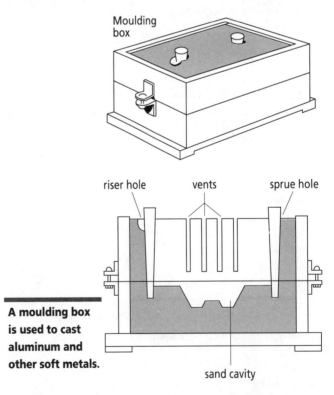

Moulding box

riser hole   vents   sprue hole

A moulding box is used to cast aluminum and other soft metals.

sand cavity

Injection moulds are used to process granular plastic. The plastic is melted and then forced into a closed mould using a great deal of pressure. You are limited in your projects by the size of the injection machine and the availability of the metal moulds.

Joinery Glue Chart

Glue	Set Time	Woods	Plastics	Metals	Ceramics
Polyvinyl emulsion (carpenter's glue)	clamp and set for 24 h	X			
Resorcinol resin (mix part A resin with part B catalyst)	clamp and set for 24 h	X			
Contact cement (apply to both surfaces)	apply and set for 30 min, then place surfaces together	X		X	
Methylene chloride (dissolves plastics to create a weld-like joint)	through capillary action, soak surface in bonding agent until plastic softens and then bond; allow to set for 24 h		X		
Mastics (heavy adhesive used to attach wallboard flooring)	method of application and set time varies according to product — read label instructions		X		X
Epoxy (mix part A resin with part B catalyst)	sets in 5–10 min, gains full strength in 24 h	X		X	X
Anaerobics (cure only when metal is present and air is excluded)	sets in 4–15 min, reaches full strength in 24 h			X	
Cyanocrylate (super glues)	sets in seconds, gains full strength in minutes	X	X	X	X
Toughened acrylic (apply part A to one surface, apply part B activator to other surface)	sets in 1–4 min, gains full strength in 4–24 h	X	X	X	X

Joining

Joining is the process of fastening two or more separate pieces of material together.

There are three general categories of joining: bonding (such as gluing, soldering, and welding), mechanical joints (such as wood joints), and mechanical fasteners (such as nails, screws, and nuts and bolts). Sometimes joining requires a combination of these methods.

Bonding

Bonding is a method of joining materials by using a substance or heat.

GLUES

There are various types of glues for bonding materials. A good-quality glue will bond for a long time. The keys to good glue joinery are making sure you have chosen the correct glue for the job and the correct clamp. Dry-clamp your work first (with no glue) to make certain that the parts fit together properly. Read and follow the directions on the glue container.

Have your teacher check the glue joints and the clamping set-up. The chart above shows the common glues you will find in the classroom, and their uses.

In addition to liquid glues, the hot glue gun can be used for bonding. The glue is waterproof and fairly strong, and it comes in stick form. The glue must be applied with the electric gun, which melts the glue.

CLAMPS

Just as clamps are important for securing your material before cutting, they are equally important in joining.

Clamps allow a glue to set and the joint to form between two or more materials by applying pressure. For a reminder of the different types of clamps, see page 154 in this chapter.

SOLDERS AND SOLDERING IRONS
Soldering is a method of joining metal. This is very similar to using a hot glue gun, but the temperature is higher and the **solder** (glue) is a metal alloy. Soldering is used extensively in the electrical industry because the solder forms a permanent bond and conducts electricity.

A solder that contains tin and lead is called a soft solder because it has a low melting point. This solder is easy to find in hardware stores. Sheet metals are bonded using soft solder. You will require the correct solder, soldering flux (a corrosive paste that cleans metal surfaces for soldering), and an electric soldering iron. The pieces to be soldered must be held together and the joint must be wiped with flux. Place the broad surface of the soldering iron on the metal and slowly draw it along the joint. Add solder to the tip of the iron by placing the solder wire against it.

Hard solder contains silver. It is used for attaching jewellery parts and is also used in the aerospace and food-processing industries. Hard solders make more permanent and stronger bonds in ferrous and non-ferrous metals. The surfaces to be soldered must be cleaned thoroughly with flux.

Testing the solder on a sample of material

WELDING TOOLS
Bonding materials by using pressure or heat is called **welding**. In gas welding, acetylene and oxygen are used to produce a heat that reaches 2800–3500 °C. Gas welding is used primarily by repair people to bond metals in a brazing or fusion process, to cut metals, and to loosen rusty nuts and bolts.

 Caution: Oxyacetylene gas is potentially explosive when mixed with surrounding air. Welding should be done with extreme caution. Wear protective clothing and welding glasses.

In arc welding, intense heat is created by a controlled short circuit across an air gap between the base metal and the electrode of the welding tool. Successful arc welding requires maintaining the rod at a specific distance from the metal, an electrode angle of 5°–20° in the weld pool

(the area being welded), a uniform electrode travel rate, and a current setting strong enough to melt the metal but not so strong as to penetrate it.

Caution: Always wear an approved arc welding helmet when looking at an arc weld.

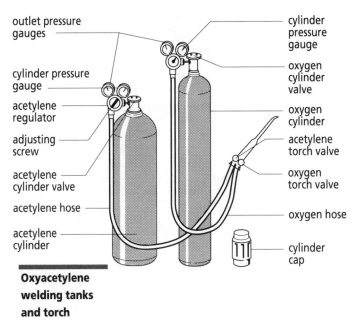

Oxyacetylene welding tanks and torch

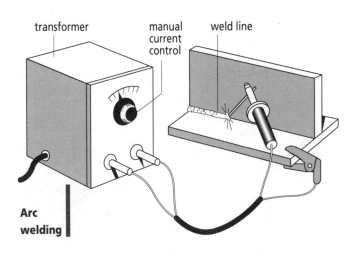

Arc welding

Mechanical Joints

Mechanical joints are joints that are cut so that they make use of a simple machine. For example, a joint may work because it is wedge-shaped. Just as the pieces of a jigsaw puzzle fit together to form one unit, so do mechanical joints. Common tools, such as saws, are used to cut these joints. Mechanical joints are used in ceramics, such as in interlocking bricks, and in plastics joinery. However, their principal use is in wood joinery. There are several types of wood joints: **dowel**, **butt**, **rabbet**, **mortise and tenon**, **half lap**, **lap**, and **dado**.

Simple joints

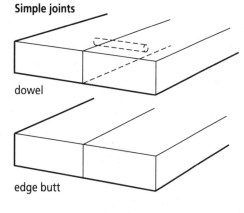

dowel

edge butt

Corner joints

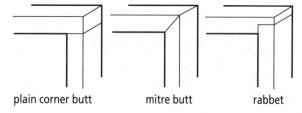

plain corner butt   mitre butt   rabbet

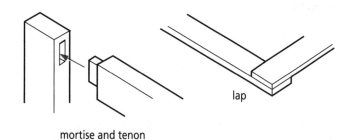

mortise and tenon

lap

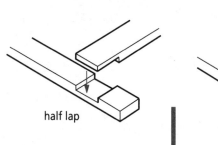

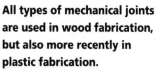

half lap   dado

All types of mechanical joints are used in wood fabrication, but also more recently in plastic fabrication.

Mechanical Fasteners

Mechanical fasteners are commonly called "hardware" because they are found in a hardware store. However, they are mechanical in that each type of fastener is a small machine. Each fastener acts as a bridge between two materials. Common types of mechanical fasteners are nails, screws, nuts and bolts, and pop rivets.

NAILS

Various types and sizes of **nails** are available. They are commonly used to join wood to wood. Mild steel nails are very common; nails made of copper, brass, stainless steel, and aluminum have special applications. Hardware stores also carry finishing nails, ring nails, spiral nails, and concrete nails.

For effective bonding, a nail should penetrate at least half the thickness of the attached wood. Hold the nail in one hand and a **claw hammer** in your other hand. First set the nail with a few light blows of the hammer. Then remove your hand and drive the nail until it is flush with the wood surface.

A nail set is a tool that hammers the head of a finishing nail below the surface of the wood. A small amount of wood fill can then be used to hide the nail head.

Hardwoods must be pre-drilled before nailing; otherwise the wood will split, marring your project. When dismantling nailed woods, be certain to remove all exposed nails. Serious injury can occur if a person steps or falls onto a protruding nail. Nails must never be used on children's toys as they pose a safety hazard. On fine furniture, the use of nails will spoil the quality of the work.

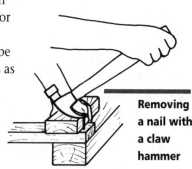

Removing a nail with a claw hammer

Caution: Safety glasses must be worn when using a hammer to drive nails, in case of flying particles of metal.

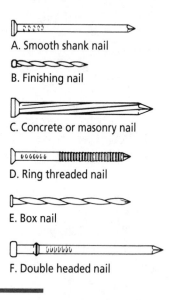

A. Smooth shank nail

B. Finishing nail

C. Concrete or masonry nail

D. Ring threaded nail

E. Box nail

F. Double headed nail

Common nails

SCREWS

A more secure way of joining wood to wood is to use **screws**. They are available in a variety of sizes and materials. Plastics and sheet metals can be joined using special screws. Wood screws are sized according to the diameter (gauge) of the shank and the length of the screw. The diameter of the shank is indicated by a gauge number 0 to 24; the most common gauges are 6 to 12. The length of a flathead screw is its overall length, whereas the lengths of the roundhead and the oval head are measured from the bottom of the head.

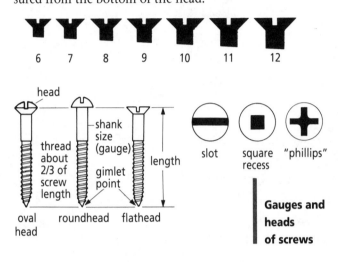

Gauges and heads of screws

You must drill a clearance hole and, if necessary, a countersink hole on the first piece of wood. The clearance hole allows the screw to fit with minimal clearance. The countersink hole allows the head of the screw to be level with the wood surface. On the second piece of wood, drill a pilot hole the size of the root of the screw. The threads on the screw cut into the sides of the pilot hole, creating an effect similar to a vise clamping a piece of wood.

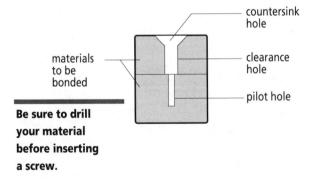

Be sure to drill your material before inserting a screw.

NUTS AND BOLTS

For large structures or sturdy benches and tables, **nuts** and **bolts** are ideal mechanical fasteners. They can be used to fasten many materials. Nuts and bolts also allow for easy dismantling of a project for storage or trans-

portation. To protect the material and to evenly distribute the pressure of a nut and bolt, use a flat washer. A hole slightly larger than the diameter of the bolt must first be drilled through the material.

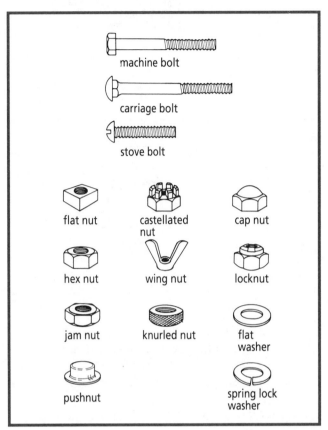

POP RIVETS

Pop rivets can be used to form an excellent mechanical bond. They are used in industry for joining metal to metal or metal to plastic. The pop rivet fits into a pre-drilled hole. By successive squeezes on the handles of the riveting pliers, the rivet expands; eventually the excess metal drops off, leaving a formed head on the surface.

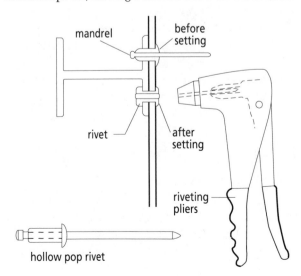

Finishing Materials

Material	Filing required	Abrasive paper or cloth	Buffing	Coating
Woods	yes	sandpaper, 80–150 grit	no	urethane, oil, wax, paint, stain, shellac
Plastics	yes	wet and dry sandpaper, 400 grit	yes	none required
Ferrous metals	yes	emery cloth, coarse to fine	no	metal paint
Non-ferrous metals	yes	emery cloth, coarse to fine	yes	metal oxides, wax
Ceramics	no	for glass, wet and dry sandpaper, 400 grit	no	none required, clay often finished with glaze or paint

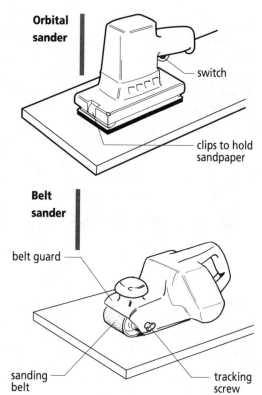

Orbital sander — switch, clips to hold sandpaper

Belt sander — belt guard, sanding belt, tracking screw

MOVING JOINERY

Types of **moving joinery**, such as hinges, are more complex mechanical fasteners. You can make a moving joint using a dowel as an axle in a wood joint. Hardware such as eye screws and fine wire can also be used to make moving wood joints. Metal rods capped with friction-fit pushnuts make excellent axles for wooden toys.

Finishing

Finishing is the final process in fabrication, before the product is used. Finishes are applied to projects to protect them and to give them a pleasing appearance. The chart at the top shows how woods, plastics, metals, and ceramics are commonly finished.

All soiled cloths and paper used in finishing work must be discarded in the proper receptacle for safety reasons. Brushes must be cleaned with a solvent and then washed with soap and warm water. Water-based, environmentally friendly finishes require only soap and water.

Finishing Tools

SANDPAPER AND SANDERS

The goal of **sanding** is to make flat, clean surfaces and edges. The various grades of sandpaper allow rough edges and surfaces to be smoothed. Sandpapers are composed of an adhesive, an abrasive, and a backing material. The adhesive is usually waterproof. The abrasive can be flint, garnet, aluminum oxide, or silicon carbide. The backing can be various grades of paper or cloth.

For wood, sand with the grain using a progressively finer grade of sandpaper. Start with the **orbital sander** (the belt sander should only be used if the wood is very hard), and then finally a sanding block. Begin sanding your work with an 80 to 100 grit paper. Use a folded piece of sandpaper to sand the edges. Move up to 120 grit paper and then to 150 grit paper. The following are typical grits of sandpaper:

Very fine	300	400	600
Fine	150	180	220
Medium	80	100	120
Coarse	40	50	60

For plastics and glass, progressively finer wet and dry sandpaper is used to remove cutting marks and to smooth edges. Metals are sanded with progressively finer grades of emery cloth and then polished.

COATINGS

Once materials are smooth, **coatings** are applied to them. Most can be applied with a paint brush or soft cloth. Care should be taken to select the coating that is most environmentally friendly. Be sure to read and follow the directions on the container for application and clean-up.

For more information on coatings, see the finishing chart at the top of this page.

Points for Review

- Fabrication is the process of building, constructing, or manufacturing according to plan.
- Know the qualities of your material, and be aware of the possible fabrication processes.
- Test each process before fabrication so you know how to incorporate each process into your project.
- List the sequence of processes on a fabrication chart.
- Measuring accurately will ensure that your design will be the size you planned, you will not waste material, and your cutting, bending, and joining will be accurate.
- After measuring your materials, process them by cutting, bending, pouring, joining, or finishing.
- Tools that cut materials include files, chisels, planes, saws, and drills.
- Electrically powered machines that cut materials include sabre saws, scroll saws, band saws, routers, and lathes.
- Tools that bend materials include hammers, moulds, and jigs.
- Joining is the process of fastening two or more separate pieces of material together.
- Bonding is the process of joining materials by using a substance or heat.
- Mechanical joints are joints that are cut so that they make use of a simple machine (for example, dowel, butt, rabbet, and mortise and tenon joints).
- Mechanical fasteners include nails, screws, nuts and bolts, and pop rivets.
- Pouring is using material in a liquid state to make a final solid shape.
- Common types of moulds are moulding boxes and injection moulds.

Terms to Remember

backsaw	crosscut saw	lathes	rasp
ball-peen hammer	cross-filing	level	rip saw
band saw	cutting	marking	routers
bar clamp	dado	marking gauge	rubber mallet
belt/disc sander	dividers	measuring	sabre saw
belt sander	dowel	measuring tape	sanding
bench vise	draw-filing	mechanical fasteners	saw pitch
bending	drill bit	mechanical joints	saws
beverly shear	drill press	metal lathe	screws
bolts	drills	micrometer	scriber
bonding	fabrication	mill	scroll saw
box and pan brake	files	mortise and tenon	snips
butt	finishing	mould	solder
callipers	hacksaw	moulding box	soldering
C-clamp	half lap	moving joinery	strip heater
centre punch	hammer	nails	table saw
chisels	hand brace	nuts	thermoforming
clamps	hand drill	orbital sander	try square
claw hammer	handscrew clamp	plane	vernier callipers
CNC router	injection moulds	pop rivets	welding
coatings	jigs	portable electric drill	wire cutters
combination square	joining	pouring	wood lathe
coping saw	lap	rabbet	

Applying Your Knowledge

1. Why is it that wood cannot be poured? What qualities do other materials have that are different from wood that allow them to be poured?
2. List the basic fabrication processes and describe them briefly.
3. Imagine that you need to design a toy car.
    a. Find a photo in a magazine of the type of car you would like to design.
    b. Make a list of the fabrication processes you think you will have to use to build a model of this car.
    c. List the tools needed for each process in part b.
4. Choose an object (such as a block of wood or a ceramic mug) and measure all its dimensions using any of the measuring tools in your classroom. Sketch the object and indicate its dimensions.
5. Explain two important uses of clamps in fabrication. Explain in a few sentences why each use of the clamp is important.
6. Think of a material you might need to fabricate a design.
    a. Write down the best way to cut the material and the corresponding tool needed.
    b. Suggest a clamp to be used while cutting.
    c. Discuss these decisions with your teacher.
7. Name the two methods of filing. What safety procedures should be taken when filing? How can you clean a file? Be sure to review your answers with your teacher.
8. Copy the following list into your notebook:
    *   sheet plastic
    *   sheet metal
    *   ceramic tile
    *   angle bar metal
    *   wire
    *   wood (along the grain)
    a. Beside the materials, list the hand tools you would use to make a straight cut in each.
    b. Beside the materials, list the electrically powered tools you would use to make a straight cut in each.
    c. Beside the materials, list the tools you would use to drill a 1 cm diameter hole in each.
9. Bring an old plastic toy from home to your class. With a classmate, do the following:
    a. Take the toy apart carefully, making sure to save all the parts.
    b. List all the different parts.
    c. For each part, list the different processes that you think were used in its fabrication.
10. List the different ways that metals can be bonded.
11. Your teacher will provide you with the following samples:
    *   a wood block
    *   a piece of aluminum
    *   a ceramic tile
    *   a piece of angle iron
    *   a piece of sheet plastic
    a. Leave these samples overnight in a can filled with water, and examine them the next day.
    b. List the samples and the effect of the water on each.
    c. Suggest a coating for those samples that you think need one.

Part 2 Projects

The following projects are ready for you to undertake. In some cases, you are given the materials and the fabrication steps. In others, more decision making is left up to you. The aim of these projects is to give you the confidence to fabricate your designs.

Making a Solar Oven

*T*his project involves cooking a pizza in a cardboard box using a renewable source of energy — the Sun. Related concepts are radiation, reflection, conduction, evapora-tion, condensation, and insulation. You may wish to do further research into these areas and design other projects that use the Sun's energy.

Materials

cardboard box, 45 cm x 45 cm x 45 cm
foil-backed foam insulation
silicone gel
caulking gun
piece of glass, 45 cm x 45 cm
duct tape
sticks

Fabrication

1. Cut the top of the box on a 30° angle.

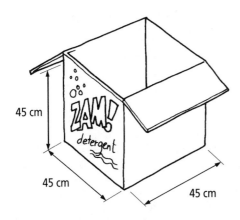

45 cm

45 cm

45 cm

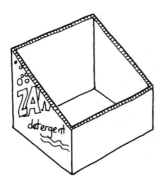

2. Line the inside of the box with the insulation (this is used on houses before the bricks are put on and can be found in building supply stores). If the insulation is not available, use tin foil.

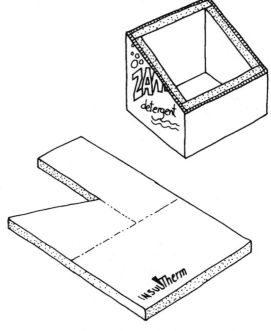

3. Use the caulking gun to seal the cracks at the edges of the insulation with silicone gel.

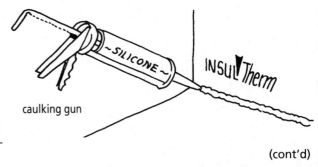

caulking gun

(cont'd)

4. Cover the top of the box with the piece of glass. Secure it with duct tape. Make sure to seal the edges well.

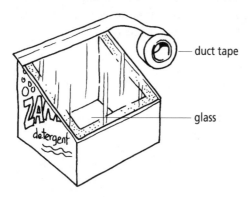

— duct tape

— glass

ZAM detergent

5. Cut a small rectangular door in the back of the oven, wide enough for a small pizza to fit through, and hinge it with a piece of tape. Glue a piece of insulation to the inside of the cardboard door.

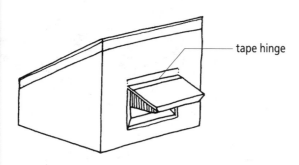

— tape hinge

6. Add three reflectors made of insulation to the outside of the box to direct the sunlight into the box. Prop up these reflectors with sticks.

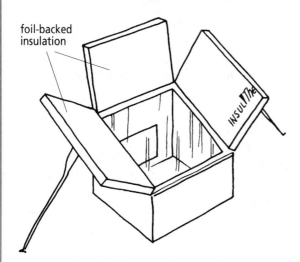

foil-backed insulation

INSULTherm

7. Slide a small frozen pizza, about 20 cm in diameter, through the door and position the oven to get the maximum amount of sunlight through the top window.

   Watch what happens. Condensation will occur and the cheese will start to melt. The aroma will come through the box as the pizza cooks. You can use an oven thermometer to record the inside temperature. Time how long it takes for the pizza to cook.

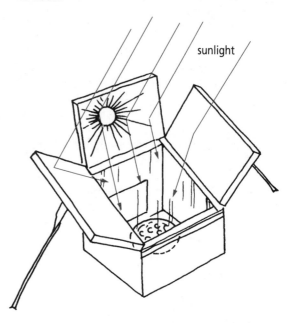

sunlight

(NOTE: Solar energy works well in the summer, but cooking a pizza this way in the winter is a challenge. In winter, make a solar shelter from three full sheets of foil-backed foam insulation. Place the solar oven in the protection and reflection of the solar shelter. The outside temperature of the box will rise to 30°C, and the inside temperature will reach 100°C and higher.)

Making a Tower

*T*his project is a good way to gain experience in fabricating a common type of structure.

Materials

base, 2 cm x 15 cm x 15 cm pine
uprights, 4 pieces 1 cm x 1 cm x 24 cm pine
horizontal members, 1 cm x 1 cm pine, cut to different lengths
diagonal members, 1 cm x 1 cm pine, cut to different lengths
hot glue gun

Fabrication

1. Make a full-size diagram for one side of the tower. This diagram will enable you to determine the length for each horizontal and diagonal member. Cut all the pieces you need for the four sides of the tower.

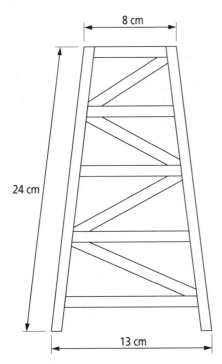

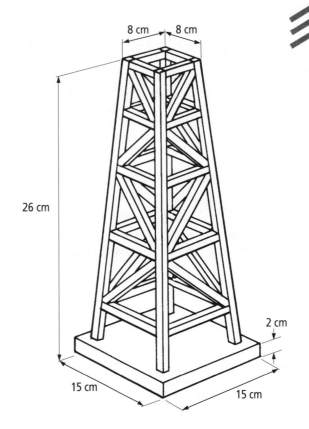

2. On a flat surface, carefully assemble two sides.

3. Hold the two sides facing each other and place the bottom horizontal members to attach the two sides.

4. Mark the positions for the uprights on the base. Glue the tower to the base.

5. Glue in place the remaining horizontal and diagonal members to complete the last two sides of the tower.

6. Test the tower's strength with heavy books.

Making an Electromagnet Crane

*I*n the manufacturing and construction industries, electromagnets are used in cranes to pick up heavy pieces of metal.

Materials
1.5 V battery
iron bolt, about 5 cm long
30 gauge coated copper wire, 1 m string
2 nails, 2.5 cm long
base, 2 cm x 20 cm x 12 cm pine

vertical post, 2 cm x 2 cm x 25 cm pine
horizontal post, 2 cm x 2 cm x 15 cm pine
diagonal support, 2 cm x 2 cm x 5 cm pine, cut at 45°
    angle at both ends
base for switch, about 2.5 cm x 5 cm x 5 cm pine
2 thin metal strips
carpenter's glue
finishing nails
tape

Fabrication

1. To make the electromagnet, wrap the bolt with about 100 turns of the copper wire. Leave enough wire at the beginning and the end to extend to the nail connectors. Cut two shorter pieces of wire to connect the nails to the switch and the battery.

2. To make the structure, drill a hole wide enough to thread the string through about 2 cm from one end of the horizontal post.

3. Assemble the vertical, horizontal, and diagonal members using glue and finishing nails. Attach the horizontal post to the base with glue and finishing nails.

4. Hammer the 2.5 cm nails into the base about 1 cm in from each side (see the diagram). Do not hammer them in all the way, so you can connect the wires to them. The insulation will have to be removed from the ends of the wire at all connections for the circuit to work.

5. Suspend the electromagnet with string from the horizontal post.

6. Make the switch by nailing 2 metal strips to the base as shown in the diagram. Bend the longer strip up; in this position, the circuit will be broken. Connect the wires to the switch with tape as shown.

7. Use tape to connect the wires to the 1.5 V battery. When the circuit is closed, the electromagnet will be charged and should pick up metals that contain iron.

8. Experiment with lifting different metals with the electromagnet.

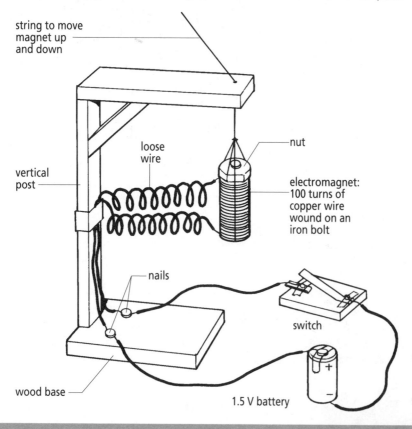

string to move magnet up and down

vertical post

loose wire

nut

electromagnet: 100 turns of copper wire wound on an iron bolt

nails

wood base

switch

1.5 V battery

Making a Small House

*W*hen a house is being designed, often a model will be built. Building a section of a small house will give you experience in fabricating this common structure.

Materials
pine strips, 0.5 cm thick x 2 cm wide
hot glue gun

Fabrication

1. Cut pine strips 15 cm long to make the walls. Cut pieces 50 cm long for the top and base plates of the house.

2. Glue 15 cm strips to a 50 cm base, spacing them 4.5 cm apart.

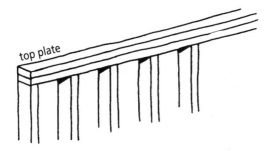

3. Add a double plate at the top.

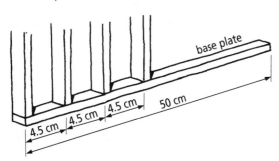

4. Cut a door and a window. Double the framing around the openings.

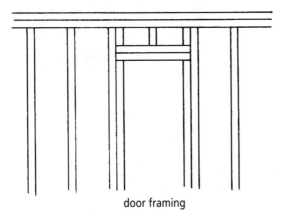

door framing

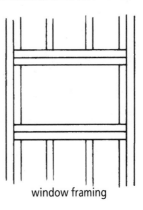

window framing

5. Add a centre wall using base and top plates that are 20 cm long.

6. Glue a 25 cm long vertical roof support to the end of the centre wall.

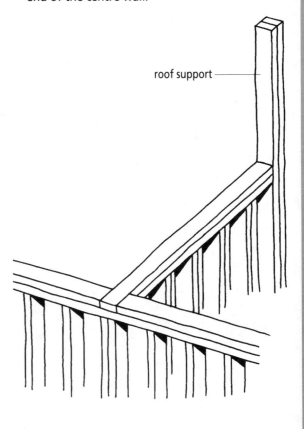

roof support

(cont'd)

7. Glue a ceiling rafter support to the vertical roof support, resting it on the top plate.

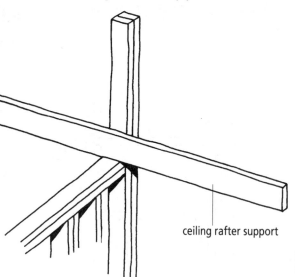

ceiling rafter support

8. Glue ceiling rafters to the ceiling rafter support, spacing them 4.5 cm apart.

ceiling rafters

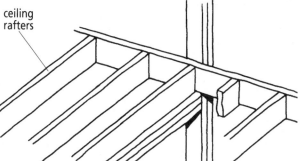

9. Glue a ridge pole to the end of the vertical roof support.

ridge pole

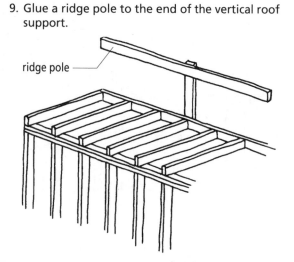

10. Glue roof rafters between the ridge pole and the ceiling rafters.

roof rafters

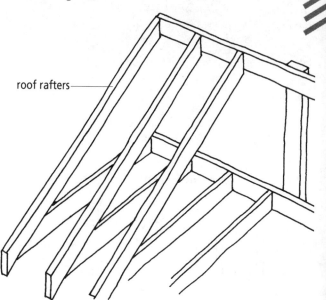

11. Attach a fascia board to the ends of the overhanging roof rafters.

fascia board

12. The completed section of the house will look like the diagram following.

(cont'd)

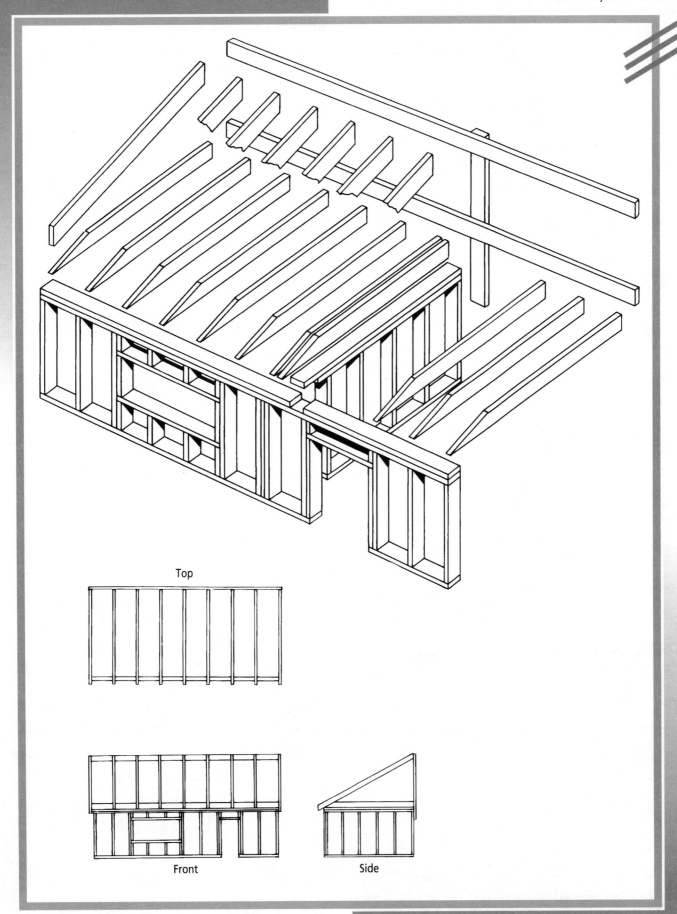

Top

Front

Side

Making a Letter Holder

Need: To create something that will organize letters on a desk.

Brainstorming: (possible ideas)
- Create a metal filing cabinet.
- Create a wooden in/out box.
- Create an in/out basket.
- Create a plastic letter holder.
- Create a cork bulletin board.

Design Brief: To design a plastic letter holder that will fit on a desk.

Researching: As a result of researching, you decide on the following shape:

letter holder
made of
bent plastic

Continue the researching and other design stages in order to fabricate your own letter holder. Feel free to modify the design in any way you want. This project will give you experience in heating and bending plastics.

Making a Key Chain

Need: To create a key chain that is a personal design.

Brainstorming: Sketch a minimum of five ideas. Use the designs shown to help you generate your own ideas. Be sure to go through the steps of the design process to create your key chain.

Making a Jewellery Holder

Need: To create something to hold jewellery that allows easy access to each piece.

Brainstorming: Sketch a minimum of five ideas. Use the designs shown to help you generate your own ideas. Be sure to go through the steps of the design process to create your jewellery holder.

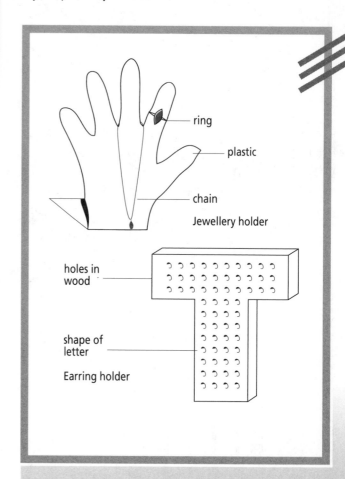

ring

plastic

chain

Jewellery holder

holes in
wood

shape of
letter

Earring holder

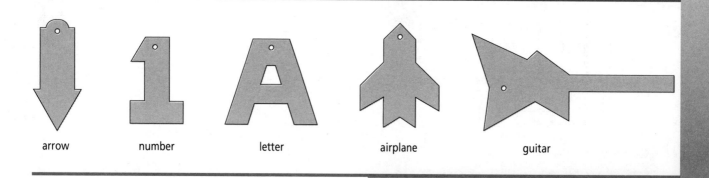

arrow   number   letter   airplane   guitar

Making an Electric Signal

You see many signals each day. For example, at a railroad crossing a signal lights up and drops a bar across the road as the train approaches.

Use the diagram to help you build your own electric signal. The information on the diagram will help you to draw up a bill of materials and a fabrication chart. Be sure to consult your teacher throughout the entire design process.

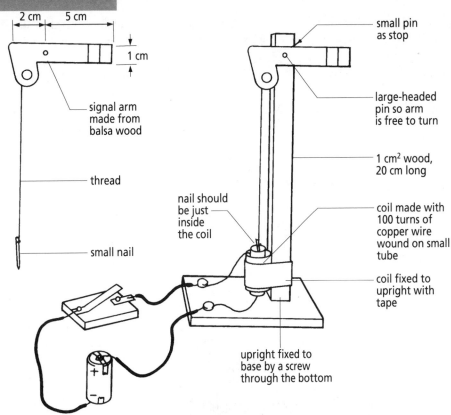

signal arm made from balsa wood

thread

small nail

2 cm   5 cm

1 cm

small pin as stop

large-headed pin so arm is free to turn

1 cm² wood, 20 cm long

coil made with 100 turns of copper wire wound on small tube

coil fixed to upright with tape

nail should be just inside the coil

upright fixed to base by a screw through the bottom

Making a Metal Box

Follow these steps to fabricate a sheet-metal box.

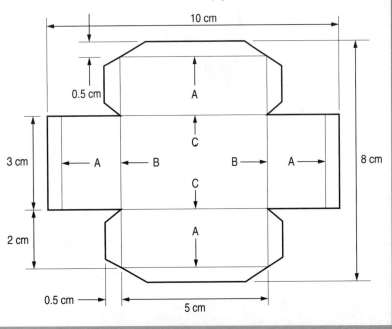

1. Ask yourself: "What could I use this box for?"

2. On paper, determine the dimensions for the box. What are the ID and the OD of the box?

3. Draw a pattern for the box. Use the diagram that follows as a guide. The measurements are only suggestions — make your box to suit the materials it will hold.

4. Use snips to square off the piece of metal to your calculated OD.

5. Carefully and exactly, transfer your pattern to the metal using a scriber and a ruler.

6. Cut out the metal.

7. Using a box and pan brake, fold all the edges labelled A to make safety edges.

8. Using the box and pan brake, fold the edges labelled B and then fold the edges labelled C.

9. Solder the corner lap joints neatly.

10 cm

0.5 cm

3 cm

2 cm

0.5 cm

5 cm

8 cm

A

C

A   B   B   A

C

A

A

0.5 cm

Making a Musical Triangle

Use the following diagrams to help you make a musical triangle, a type of percussion instrument.

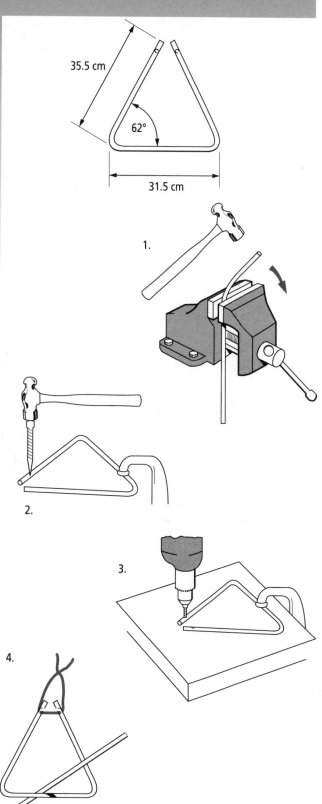

Making a Musical Instrument

Need: To create a simple musical instrument to be used in music class.

Brainstorming: Sketch a minimum of five ideas. Use the following designs to help you generate your own ideas. Be sure to go through the steps of the design process to create your musical instrument.

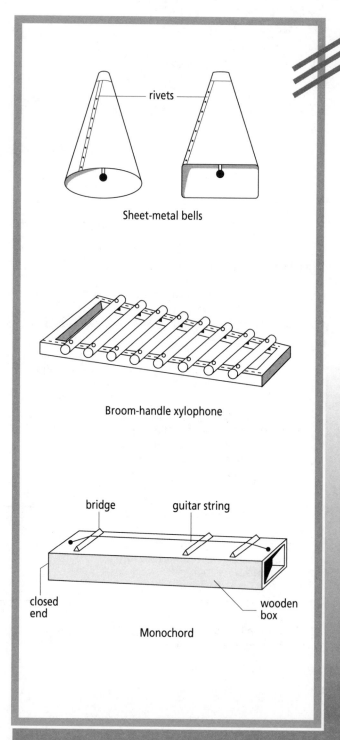

Making a Book Holder

Need: To create something to hold the books in the living room at home.

Brainstorming: Sketch a minimum of five ideas. Use the following designs to help you generate your own ideas. Be sure to go through the steps of the design process to create your book holder.

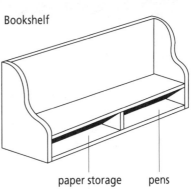

Bookshelf

paper storage   pens

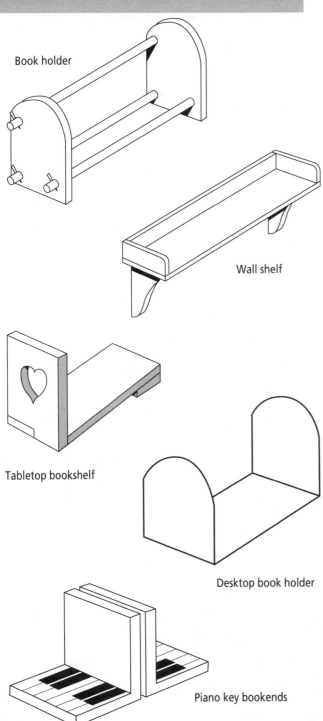

Book holder

Wall shelf

Tabletop bookshelf

Desktop book holder

Piano key bookends

Making Custom Jewellery

Need: To create a piece of jewellery to give to a friend as a present.

Brainstorming: Sketch a minimum of five ideas. Use the following designs to help you generate your own ideas. Be sure to go through the steps of the design process to create your jewellery.

Use studs or loops to fasten to ears.

earring shapes of wood or plastic

Glue to saftey pins or jewellery findings.

pin shapes of wood or plastic

Making a Pen Set

Design your own pen set. You can use acrylic or wood to make the base. Your teacher will provide the pen and the pen funnel. Sketch five possible base shapes and then choose the design that is best for you. Use the diagrams below to help you fabricate your pen set, first making sure you go through the steps of the design process.

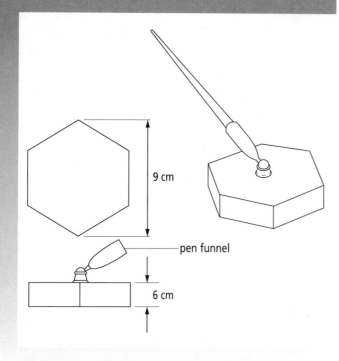

9 cm

pen funnel

6 cm

Making a Magazine Holder

Need: To create a holder for magazines to keep them organized in your room at home.

Brainstorming: Choose one of the following designs for your magazine holder. Feel free to modify any of these designs. Be sure to go through the steps of the design process to fabricate your magazine holder.

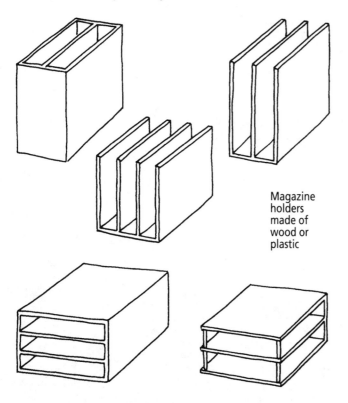

Magazine holders made of wood or plastic

Making a Utensil Holder

Need: To create something that will hold a spatula, a soup spoon, tongs, and a large mixing spoon in the kitchen in your home.

Use the design process to fulfill this need. Be sure to go through all the steps of the design process, consulting your teacher at the various stages.

Design Solutions — Starting from Scratch
The following are some situations that require design solutions. Read the situations carefully and then use the design process described in Chapter 2 to find solutions. Be sure to consult your teacher after each step, especially before you begin fabrication.

1. Practically all finished goods must be shipped in some type of container built specifically for them. Containers are a special type of puzzle; when completed, they make a three-dimensional shape. Making containers will help you learn more about structure.

   Use these diagrams as ideas in order to design and fabricate a container to hold a breakable object (for example, a ceramic mug). Your container should have opening and closing flaps on each end.

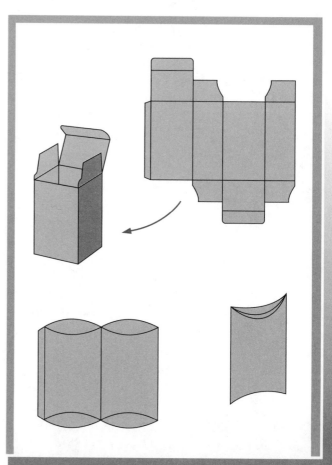

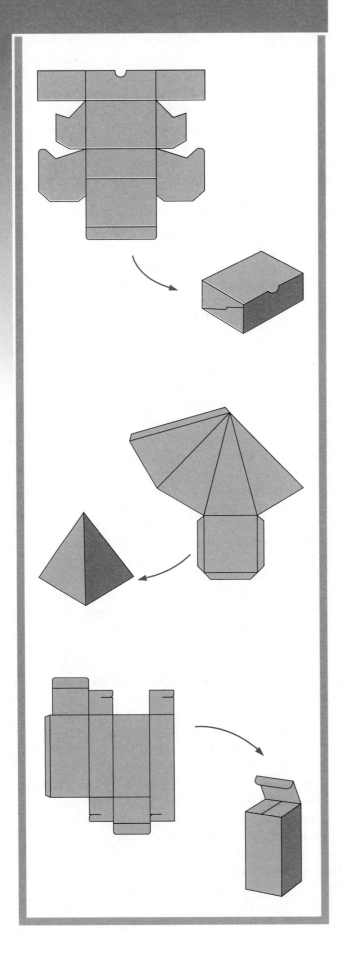

2. Take apart a wind-up toy car. If you cannot find an old one at home, your teacher will provide one. Most wind-up cars have a mechanism like the one below.

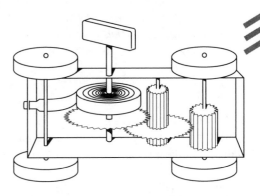

Be sure to take the car apart carefully and save all the pieces. Sketch the parts and note what each is used for. This research will help you to design your own wind-up toy. Be sure to go through all the steps of the design process to create your wind-up toy, consulting your teacher at the various stages.

Photo/Illustration Credits

p. 1 Courtesy of Ford Motor Company; p. 3 (bottom left) This symbol is a registered trademark of the Canadian Standards Association and has been reproduced with permission of that association; p. 4 (right, both photos) Designed by Paradox Design; p. 5 (top left) Designed by Paradox Design; (bottom right) Photo courtesy Lee Valley Tools Limited; p. 6 *Workplace Hazardous Materials Information System (WHMIS): A Guide to the Legislation.* Ontario Ministry of Labour; p. 7 (top left) Reproduced from the "Chemical Control Program Guide," Industrial Accident Prevention Association; (top right) Fendall Company; p. 8 (right) Canada Wide/Mike Peake; p. 9 (top left) Photo courtesy of 3M Canada Inc., Occupational Health and Environmental Safety; (bottom left) *Workplace Hazardous Materials Information System (WHMIS): A Guide to the Legislation.* Ontario Ministry of Labour; p. 12 (left, both photos) Courtesy of Ford Motor Company; (right) Copyright *The London Free Press*. Photograph by Susan Bradnam; p. 18 (both photos) Beverly Dywan; p. 19 (bottom right) Janet Collins, Canadian Library Association; p. 28 (left, all photos) Photos: K. Browning; p. 29 (top right) Jonathan Crinion. President: Crinion Associates; (bottom right) Gazelle Chair. Design: Crinion Associates. Photo: Shin Sugino; p. 30 (top left) Oscillating Wall Fan. Design: Crinion Associates. Photo: Doug Hall; p. 35 (bottom right) Drafix® Windows CAD™ 2.0 from Foresight Resources Corp.; p. 39 (left, both photos) The Bettmann Archive; (right, all photos) *Aviation & Aerospace Magazine*; p. 40 (left) Courtesy: P. Lees/J. Evans; p. 47 (left, all photos) NASA; (right) Courtesy — Environment Canada — Parks Service, Alexander Graham Bell National Historical Site; p. 49 (chart) Used with the permission of the Canadian Hearing Society; p. 52 (left) Plastic drums (1984) K. Browning; (right) Performance artist K. Browning; p. 54 (top right) Student Ted Monchesky holding guitar. Photo: K. Browning; p. 60 (right) Photo: K. Browning; p. 63 © CN Tower; p. 64 (bottom left) Canada Wide; (bottom right) © CN Tower; p. 65 Courtesy: Jonathan C. Evans; p. 66 (top left) The Bettmann Archive; (top right) Compliments of Lado Musical Inc.; p. 67 (centre left) Courtesy: Jonathan C. Evans; (bottom left) Photo: K. Browning; (right, both photos) Italian Government Travel Office, 1 Place Ville Marie, Suite 1914, Montreal, Quebec, H3B 3M9; p. 68 (top, all photos) Courtesy: Jonathan C. Evans; p. 69 (top right) Province of British Columbia; (bottom right) Courtesy: Jonathan C. Evans; p. 70 Allan Baker, Industrial Designer, A.O.C.A., A.C.I.D.; p. 72 (left) National Archives of Canada/PA 56401; p. 74 (top right) Courtesy of the Dominion Bridge Company Limited/National Archives of Canada/PA 135835; (bottom right) World Trade Centre; p. 75 (centre right) Italian Government Travel Office, 1 Place Ville Marie, Suite 1914, Montreal, Quebec, H3B 3M9; (bottom right) Courtesy: Jonathan C. Evans; p. 76 (left, all photos) Courtesy: Jonathan C. Evans; (top right) Photo: K. Browning; p. 82 (left) Courtesy of *MTHL News*. Photo: D. Magee; (top right) Canada Wide/Mike Peake; p. 88 (bottom left) Solec International; (top right) The Bettmann Archive; p. 89 (left, both photos) Reuters/Bettmann; p. 90 (bottom) Ontario Hydro; p. 92 (top left) The Bettmann Archive; (bottom left) Courtesy: Jonathan C. Evans; p. 93 (centre left) The Bettmann Archive; p. 94 (left, both photos) Photos courtesy of Sony of Canada Ltd. Sony is a registered trademark of Sony Corporation Tokyo, Japan; p. 95 (right) National Research Council of Canada; p. 99 Photo supplied courtesy of Hewlett-Packard Company; p. 103 (centre left) National Archives of Canada/RD 664; (bottom left) Courtesy of Otto Bock Industries of Canada Limited; (top right) National Archives of Canada/PA 119899; (centre right, top) National Archives of Canada/PA 61677; (centre right, bottom) National Archives of Canada/PA 10539; (bottom right) Photo courtesy Ford of Canada; p. 104 (top left) Photo courtesy Boeing Defense & Space Group; (bottom left) Courtesy of Conklin & Garrett Ltd.; (right) NASA; p. 107 (bottom left) Department of Fisheries and Oceans, Vancouver, British Columbia; p. 109 (centre left) Courtesy of Health & Welfare Canada; p. 110 (top right) Courtesy: Jonathan C. Evans; p. 112 (centre left) Photo courtesy of Caterpillar Inc.; p. 114 (bottom left) Neg./Trans. no. 325985. Photo by Anderson. Courtesy Department of Library Services, American Museum of Natural History; p. 115 (bottom left) Photo courtesy of Caterpillar Inc.; p. 117 (top) Drawings produced by Spar Aerospace Limited. Reproduced by permission of the Canadian Space Agency; p. 121 (centre right) Photo courtesy Ford of Canada; p. 127 (left) Metropolitan Toronto Department of Works; p. 128 Photo by Alan Rosenthal; p. 130 B.C. Forest Service; p. 132 (right, both diagrams) B.C. Forest Service; p. 133 (top) Canadian Wood Council; p. 136 (left, both photos) The Moving Store — Designed by Susan Ramsay, President and Founder of Rent A Boxx Moving Systems, Inc. and The Moving Store; (top right) Photo of acrylic awards courtesy of Clearmount Plastics Limited; (centre right, top) Vinyl binders shown are manufactured by E.B. Looseleaf Inc., Toronto, Ontario, Canada. A wide range of vinyl stationery products are also produced; (centre right, bottom) Ride 'em Amusements; (bottom right) Product photo courtesy of Rubbermaid Canada Inc.; p. 138 (top) Photo courtesy of Ford Motor Company of Canada Limited; p. 139 (top right) Photo courtesy of the Canadian Copper & Brass Development Association; (centre right) Noranda Minerals Inc.; p. 140 (top left) Noranda Minerals Inc.; (bottom left) Photo courtesy of Dofasco Inc., Hamilton, Ontario, Canada;

p. 142 (both photos) Alcan Aluminum Ltd.; **p. 143** (top left) On loan to The George R. Gardiner Museum of Ceramic Art; (centre left) Photo courtesy of Terra Cotta Ceramic Supplies Ltd., Scarborough, Ontario. Suppliers of hobby ceramics and art and craft materials; (bottom left) CORNING WARE® Glass-Ceramic Cookware. Courtesy of Corning Canada Inc.; (top right) Toronto Eaton Centre, supplied by Acme Slate & Tile Co. Ltd.; (centre right) Courtesy of IBM Corp.; **p. 150** Photo: Claus Andersen; **p. 152** (top right) and **p. 154** (bottom left) From *Exploring Industrial Arts* by Brian L. Henderson. Used by permission of McGraw-Hill Ryerson Limited; **p. 155** (top left and centre right) From *Exploring Industrial Arts* by Brian L. Henderson. Used by permission of McGraw-Hill Ryerson Limited; (bottom right) Stanley Tools/Division of Stanley Canada Inc., 1100 Corporate Drive, Burlington, Ontario L7L 5R6; **p. 156** (top and centre left) Stanley Tools/Division of Stanley Canada Inc., 1100 Corporate Drive, Burlington, Ontario L7L 5R6; **p. 157** (bottom left and centre right) Courtesy of Delta International Machinery, Guelph, Ontario; **p. 158** (top and bottom left) Courtesy of Delta International Machinery, Guelph, Ontario; (top right) From *Exploring Industrial Arts* by Brian L. Henderson. Used by permission of McGraw-Hill Ryerson Limited; (centre right) Stanley Tools/Division of Stanley Canada Inc., 1100 Corporate Drive, Burlington, Ontario L7L 5R6; **p. 159** (top left) Black & Decker Canada Inc.; (bottom left) Courtesy of Delta International Machinery, Guelph, Ontario; **p. 160** (top) From *Exploring Industrial Arts* by Brian L. Henderson. Used by permission of McGraw-Hill Ryerson Limited; (bottom) Courtesy of Delta International Machinery, Guelph, Ontario; **p. 161** (top) From *Exploring Industrial Arts* by Brian L. Henderson. Used by permission of McGraw-Hill Ryerson Limited; (bottom) Courtesy of Jacques Giard; **p. 165** (centre left), **p. 166** (centre and bottom right), and **p. 167** (top left and top right) From *Exploring Industrial Arts* by   Brian L. Henderson. Used by permission of McGraw-Hill Ryerson Limited.

Glossary

Abstract   To develop in unexpected ways.

Acoustic   Musical instruments that do not need an amplifier in order to be heard.

Acrylic   A thermoplastic that is strong, rigid, and impact-resistant.

Acrylonitrile Butadiene Styrene (ABS)   A thermoplastic that comes in sheet form that can be vacuum formed, pressure formed, or cold stamped.

Adapt   To use another's idea in a new way.

Aerospace   Technology that deals with flight and space travel.

Alloys   Metals that are made of different metals mixed together or metals mixed with other substances.

Alternating Current (AC)   Electric current that constantly reverses direction. Alternating current is the current in a wall outlet.

Amplitude   The maximum distance that a sound wave particle travels in one oscillation. The higher the amplitude, the louder the sound.

Annual Rings   The rings in a tree that are created by layers of spring and summer growth. You can find the age of a tree by counting the number of rings in a cross section.

Antinode   The point of a sound wave at its peak. It is the point at which there is the most vibration.

Archimedes' Screw   A broad, threaded screw encased in a cylinder that is used to raise water from one level to another.

Asymmetrical Balance   A use of space where two sides of a design are not the same.

Auger   A broad, threaded screw that is used to transport materials to different levels or along a horizontal plane. An auger bit is used to bore holes through wood.

Backsaw   A saw that is used for making straight cuts across the end grain of wood and for joinery work.

Balance   A way to clarify space. Balance can be either symmetrical or asymmetrical.

Ball-Peen Hammer   A hammer with a head that is ball-shaped on one end and flat on the other. This kind of hammer is used to bend metals and drive rivets.

Band Saw   A stationary power saw whose blade is a continuous band.

Bar Clamp   A clamp that is used to hold large-surface and frame work. It has coarse and fine adjustments for setting a moving jaw against a material.

Bark   The outer protective coating of a tree.

Bar Linkages   Linkages that use bars connected to one another by joints.

Batteries   Cells that store chemical energy.

Beams   I-shaped pieces of steel that are used in road and building construction.

Belt Drive   A form of power transmission that uses a belt to transmit force from one wheel to another.

Belt Sander   A rotary sander that uses a sanding belt to do the work.

Bench Vise   A clamp that is mounted on a workbench and that has two moving jaws to hold materials.

Bending   Changing the original shape of a material to achieve a new form.

Beverly Shear   A large, table-mounted snip that is used for cutting heavy sheet metals.

Bill of Materials   A list of the number and size of materials, tools, and hardware needed to fabricate an object.

Bolts   Threaded mechanical fasteners that work much like screws, except that they are screwed into nuts to hold them in place.

Bonding   Joining materials by using a substance or heat.

Box and Pan Brake   A table-mounted jig with a lever, which is used to bend metal.

Brainstorming   A quick means of generating ideas verbally or by writing or sketching.

Brake Calliper   On a bicycle, the part of the brake that holds the brake pads and moves them against the wheel when the brake lever is squeezed.

Bricks   Blocks formed from fired clay that are used in construction.

Bridle   The piece of string that is attached to a kite to hold it in a position that will cause the wind to push up on it.

Butt Joint   A mechanical joint that consists of two parts that abut each other.

Cable Linkages   Linkages that are made up of cables connected by joints.

Callipers   Tools that are used to measure the inside or outside dimensions of round surfaces or holes.

Cam   A specialized wheel, often heart-shaped, that is mounted on a shaft to give motion to a follower rod.

Cambium   The layer of a tree where growth takes place.

Canadian Standards Association (CSA)   This association tests and certifies products used in the marketplace. The CSA symbol indicates that a product has met certain safety and performance criteria.

Capacitor   An electronic device that stores electricity like a battery.

C-clamp   A clamp shaped like a C that is made of forged steel. This type of clamp can hold work of different thicknesses and widths.

Cement   A ceramic material that is the result of calcining limestone and red shale.

Centre Punch   A pointed metal tool that is used to mark the centre of a hole to be drilled.

Ceramics   Materials that are formed from silicon, a non-metallic compound.

Chain Drive   A form of power transmission that uses a chain to transmit force from one toothed wheel to another.

Chemical Energy   Energy that is released when heat or electric current is introduced to chemical substances.

Chisels   Cutting tools that have a sharp-edged blade, used for accurate cutting and shaping of materials.

Circuit   The path along which electricity flows.

Clamps   Tools that hold a material still against a surface or against another material.

Claw Hammer   A hammer that has two claws on one end of its head for pulling nails out of a material.

Clay   A ceramic material that is a mixture of a very fine type of sand and water.

CNC Router   A router that is controlled by a computer. CNC stands for computer numerical control.

Coatings   Finishes, such as varnish or paint, that are applied to projects.

Colour   A sensation produced when light waves strike the eye. We see colour when light waves of specific lengths are reflected or absorbed by pigments.

Combination Square   A tool that is used for squaring the end of a material, measuring several points on a material, and drawing 45° and 90° angles.

Combustion   The burning of a material using oxygen.

Compression   A pressing force on an object.

Computer-Aided Design (CAD)   A computer program that is used for making extremely detailed drawings of objects.

Concrete   A mixture of cement, sand, gravel, and water that sets to a rigid state and is then cured. Concrete is used extensively in construction.

Conduction   The transfer of heat from one particle to another.

Conductors   Materials that allow electricity to flow through them.

Consultation   The process of asking for advice from another person.

Control   A switch or other device that can stop or start the flow of electricity.

Control Surfaces   Surfaces on an aircraft that react to the flow of air and cause the aircraft to have predictable movement.

Coping Saw   A hand saw that cuts irregular shapes and curves in wood.

Covering   In a kite, the outer shell of fabric or paper that covers the frame.

Crank   A bent lever that is used to transmit motion.

Crankarm and Lever   A reciprocating drive that transfers rotary motion into linear motion.

Crankshaft   A shaft that has more than one crank, as in a car motor.

Crankwheel and Lever   A reciprocating drive that uses a wheel with a connecting rod attached to it off-centre. The rotation of the connecting rod is transferred into a linear motion.

Crosscut Saw   A hand saw that is used to cut across the grain of wood.

Cross-Filing   The technique of holding a file horizontally and drawing it along the length of a material to remove excess material.

Current   The stream of electrical charges that flow along wires, creating electricity.

Cutting   Piercing through or shearing off part of a material to achieve an overall shape.

Dado Joint   A mechanical joint in which the end of a part fits into a groove cut in another part.

Design Brief   A short statement explaining what a designer wants to fabricate.

Design Elements   Characteristics that have a direct influence on the appearance and appeal of products: shape, space, colour, and texture.

Designer   A person who develops a new or better way of creating something, either for him or herself or for another person or market.

Design Factors   Considerations that must be taken into account in planning any design: safety, function, aesthetics, ergonomics, time, materials, fabrication, and environment.

Design Process   A method that will help you when designing and fabricating objects. The steps of the design process include: (1) defining the need, (2) brainstorming ideas, (3) writing a design brief, (4) researching information, (5) planning the materials, tools, and procedures, (6) fabricating the object, and (7) evaluating the final design.

Detail Dimensions   Measurements that indicate the sizes of any contours or details of an object, other than its overall length, width, and height.

Deterioration   Environmental wear on a material.

Diagonal   An oblique, or slanting, line.

Dimension   A measurement of length, width, or height.

Dimension Lines   In a drawing, thin black lines that span the distance between pairs of extension lines and indicate measurements.

Diode   A semiconductor that allows current to flow in one direction only.

Direct Current (DC)   An electric current that always flows in one direction. Direct current is the current produced by a battery or solar cell.

Disc Sander   A rotary sander that uses a sanding disc to do the work.

Dividers   A set of dividers looks like a compass with two points, which are used to obtain a measurement and scribe it into a material.

Dowel Joint   A mechanical joint that uses dowels fitted into holes to join two pieces of material.

Draft   A precise and detailed drawing of a design.

Drafting   A technical, accurate type of drawing.

Drag   The resistance of air to the motion of an aircraft.

Draw-Filing   The technique of holding a file diagonally and drawing it along a material to produce a smooth, flat surface.

Dressing   The process of making a surface smooth. Lumber is dressed using a plane.

Drill Bit   The tip of the drill bit has the cutting edge that rotates to cut into a material.

Drill Press   A stationary power drill that drills holes accurately in different materials.

Drills   Tools that make holes in materials with a rotating drill bit.

Dynamic Load   Anything that is placed on a structure that is not part of the structure.

Effort   An applied force.

Elastic Bands   These are sources of potential energy when they are twisted or stretched and then allowed to untwist or relax.

Electrical Energy   Energy that is formed when electrons move through a material, creating a current.

Electricity   Electrical energy that flows along a path. Electricity is also the study of the generation, distribution, and use of electrical energy.

Electronics   The branch of engineering and technology that deals with the design and manufacture of devices that use electricity.

Electrons   Moving particles that surround the nucleus of an atom.

Eliminate   To delete an idea.

Ellipses   Oval shapes that have two identical ends.

Energy   The capacity for doing work; the force that makes things move.

Epoxy Resin   A thermoset plastic that has great strength and superior adhesive qualities. Epoxy resin is used in glues and in protective coatings, such as paints used in the aerospace industry.

Ergonomics   The relationship of an object to human size and form.

Evaluation   The final stage of the design process, when the fabricated object is tested in order to evaluate its success. Models, sketches, drafts, notes, charts, and the fabricated object are presented to a small or large group for evaluation.

Extension Lines   In a drawing, thin black lines that extend from the lines that define the object and that are used to indicate measurements.

Fabrication   Using materials, tools, and procedures to build, construct, or manufacture an object that will fulfill a design need.

Fabricators   People who construct products if the designers do not construct the products themselves.

Fatigue   A deterioration of metal that occurs when it is subjected to continuous application and removal of a force.

Ferrous Metals   Metals and alloys that contain iron.

Files   Tools that are used to remove excess material or to finish a surface. Files come in different shapes to file surfaces that have the same shapes.

Finishing   The final process of fabrication, when a finish such as paint or stain is applied to a project to protect it and give it a pleasing appearance.

Finishing Area   A dust-free area for applying finishes such as stain, paint, shellac, or urethane to a project.

First-Class Lever   A lever in which the fulcrum is placed between the effort and the load.

Flanges   Flat pieces of metal that are welded or riveted to the long edges of the plate of a girder to strengthen it.

Follower Rod   A rod that transmits the motion from a turning cam to do some work.

Footing   The cement base of a house foundation.

Force   A push on a structure from any direction.

Form   What a product looks like. A form is also a shape that has three dimensions: length, width, and height.

Fossil Fuels   Sources of energy that originate from decayed and compressed plant and animal matter from thousands of years ago. Fossil fuels include coal, petroleum, and natural gas.

Foundation   The base of a house, which is built on the footing and extends above the ground level.

Fractional Distillation   The process of refining crude oil by using heat to separate each of its chemical components.

Frame   In a kite, the support over which the outer covering is stretched.

Framing Line   In a kite, the cord that runs around the frame, over which the covering is stretched.

Frequency   The number of vibrations of particles in a sound wave in a second. Frequency is usually read in cycles per second.

Friction   The resistance to motion between two surfaces that touch.

Fulcrum   The fixed point on which a lever turns.

Function   What an object is used for.

Gears   Toothed wheels that are used to transmit force from one part of a machine to another.

Generator   A device that creates electrical energy.

Girders   Long, flat, rectangular plates of steel that are used in the construction of bridges and buildings.

Glass   A clear, hard, ceramic material that is formed of sand, limestone, soda ash, potash, and other minerals fused together by high heat in a kiln.

Grain   The structure of lengthwise cells in a tree, created by the contrast between spring growth and summer growth. The grain is the pattern on the face of a board.

Gravitational Energy   The energy that is created by a falling object.

Greenhouse Effect   The effect of the Earth's atmosphere trapping some heat from the Sun. When too much heat is trapped, the entire climate system of the Earth is affected.

Guy Wires   Long, strong cables that are used to support towers.

Hacksaw   A hand saw that is used to cut plastic or metal.

Half Lap Joint   A mechanical joint that uses a dado cut into two parts so that they fit into each other.

Hammer   A hand tool that is used to bend materials or drive nails.

Hand Brace   A manual drill that is used to drill large holes.

Hand Drill   A manual drill that is used to bore holes in materials that are not too hard.

Handscrew Clamp   A wooden clamp that can hold irregular angles. Each jaw of the clamp is adjusted by a separate handle.

Hardboard   A board made of wood fibres that are bonded with an adhesive and subjected to great pressure at very high temperatures.

Hardwood   Wood that usually comes from deciduous trees.

Heartwood   The inner part of a tree.

Horizon Line   The line at eye level on the horizon.

Horizontal   A line that is parallel to the horizon.

Hydraulics   An energy source that uses compressed water or other liquids to transfer motion.

Hydro-Electric Energy   Electrical energy that is produced by a generator turned by falling water.

Idea Book   A scrapbook of designs, magazine articles, notes, and sketches. An idea book can help you develop and edit ideas for design situations.

Inclined Plane   A simple machine that provides a sloping surface to gain access to lower and higher levels.

Industrial Accident Prevention Association (IAPA)   A government agency that ensures safe practices in the workplace for practically all activities. The IAPA also distributes information on treating accident victims and trains safety leaders for each workplace.

Injection Moulds   Moulds that are used to shape melted granular plastic.

Integrated Circuit (IC)   Also called a microchip, this is an electronic device that contains thousands of miniature electronic components in a thin piece of silicon seldom larger than a match head.

Intuition   A "gut feeling" about a decision; what you feel is right.

Isometric Projection   A drawing that shows the length, width, and height of a three-dimensional object.

Jigs   Devices that are used to hold wood, metal, or plastic so that they take on a particular shape.

Joining   The process of fastening two or more separate pieces of material together.

Joists   The floor beams of a house.

Kinetic Energy   The energy of an object in motion.

Lap Joint   A mechanical joint in which two joined parts cross over each other.

Lathes   Stationary power tools that use fast rotary action to spin materials as they are being cut. Lathes are used to shape cylindrical objects, such as bowls, chair legs, and axles.

Level   A tool that is used to determine whether a surface is perfectly horizontal or vertical.

Lever   A simple machine consisting of a bar supported and turning on a fixed point called the fulcrum. The lever magnifies the effect of a small effort to move a large load.

Lift   The upward force of air pressure against the bottom of an aircraft's wings.

Light-Dependent Resistor (LDR)   A type of resistor that turns on or off depending on how much light is striking it.

Light-Emitting Diode (LED)   A diode that emits light when a current flows through it.

Linear Motion   A motion that goes back and forth.

Linkages   Cams, levers, cables, cranks, and bars that are connected to one another to make up the mechanisms of machines.

Loads   In engineering, forces that push on a structure. In electricity, devices that are turned on by electric current, such as light bulbs. In lever set-ups, the masses that are moved or lifted.

Lumber   Boards that are cut from logs.

Machines   Devices made of fixed and moving parts, which do specific jobs.

Magnetic Energy   Motion that is created by the attraction and repulsion of magnetically charged materials.

Magnify   To make an idea larger.

Marking   Drawing a line or dot on the surface of a material after measuring the size needed.

Marking Gauge   A tool that scribes lines parallel to an edge into a material. The gauge has a movable head that can be adjusted to different widths.

Materials   Substances that objects are made of.

Measuring   Finding the size of an object by using a measuring instrument such as a ruler, a tape measure, or a micrometer.

Measuring Tape   A long, flexible type of ruler.

Mechanical Advantage   The ability to move a large load using a small effort.

Mechanical Fasteners   Also called hardware, these are devices that are used to join two materials. Nails, screws, nuts, bolts, and rivets are common mechanical fasteners.

Mechanical Joints   Joints that use specific cuts in materials to fit parts together, especially in wood joinery. These cuts make use of a simple machine, such as the wedge.

Medium   A solid, liquid, or gas through which sound waves can travel.

Metal Lathe   A lathe that is used to shape metals, plastics, and woods.

Metals   Materials that are made from ores, the metallic minerals in rocks.

Micrometer   A tool that combines callipers and a ruler to measure an object with great precision.

Mill   A stationary power tool that holds various cutting bits to shape woods, plastics, and metals.

Minimize   To make smaller.

Model   A small version of an object that is created to help work out any problems that may arise while fabricating the final object.

Modify   To change in order to create something new.

Moisture Content   The amount or percentage of water found in lumber.

Moment   In a lever set-up, the product of the effort multiplied by the distance from the fulcrum.

Mortise and Tenon Joint   A mechanical joint wherein one part (the tenon) fits into a recess (the mortise) cut in another part.

Mould   A hollow shape that is used for shaping metal or plastic.

Moulding Box   A box that is used for moulding metals that are in a liquid state.

Moving Joinery   Mechanical fasteners, such as hinges, that allow movement.

Nails   Slender pieces of metal that are pointed at one end. They are driven through pieces of wood to join them to one another.

Natural Gas   A fossil fuel and one of the gases that make up crude oil.

Need   The design requirement that a designer works to fulfill.

Node   The original position of a sound. A node is the point at which there is no vibration; it is the sound wave or vibration at rest.

Noise   Sound without tone, beat, or rhythm. Noise is disorganized sound.

Non-ferrous Metals   Metals or alloys that contain little or no iron.

Non-renewable   Resources that are non-renewable cannot be replaced once they are used.

Nuclear Energy   The potential energy that is stored in the nucleus of an atom.

Nuclear Fission   The process that releases the energy of an atom. To accomplish nuclear fission, the nucleus of one atom is split. The energy that is released splits the nucleus of other atoms and creates a chain of reactions and collisions.

Nuts   Threaded mechanical fasteners that are screwed onto bolts to hold them in place.

Orbital Sander   A portable power sander that uses a rotary sanding action.

Ores   Metallic minerals in rocks.

Orthographic Projection   A drawing that shows the front, top, and side views of an object separately.

Oscillation   The movement of particles in sound waves from a position of rest to another position and then back to their original position.

Overall Dimensions   Measurements that indicate the overall length, width, and height of a three-dimensional object.

Pantograph   An X-shaped bar linkage that is used to reduce or enlarge drawings.

Parallel Circuit   An electric circuit in which each control or load can operate independently from any others.

Particle Board   A board made of wood chips, splinters, flakes, and screened sawdust that are finely milled to a uniform size and bonded with glue or resin.

Perspective   The effect of distance on the appearance of objects.

Photovoltaic Cells   Solar cells, which are thin wafers of silicon that convert sunlight into electrical energy.

Physical Qualities   The qualities that define the shape and structure of a material or an object.

Pier   A support made of brick, concrete, or stone that holds up the centre of a bridge.

Pitch   In aerospace technology, the action of an airplane suddenly turning upward or diving downward. In sound, how high or low a note sounds; the greater the frequency of a vibration, the higher the pitch of the sound. Pitch is also the term for the distance between the screw threads on a screw; it equals the distance the screw travels when it is turned once.

Plane   A flat surface. A plane is also a tool that is used to square and shape wood by means of a chisel mounted in a rigid handle.

Planning   In the design process, the stage of deciding on materials, tools, and procedures to follow in creating an object, as well as considering time and cost restrictions. This stage may include preparing a bill of materials and a fabrication chart that details the steps to follow when creating the object.

Plaster   A ceramic material that is a mixture of lime or plaster of Paris (calcined limestone or gypsum) to which may be added sand or other materials.

Plastics   Synthetic materials that are composed of the hydrocarbons found in petroleum.

Ply   One of the thin layers of wood that are glued together to make plywood.

Plywood   A board made of several thin layers of wood glued together.

Pneumatics   An energy source that uses compressed air or other gases to transfer motion.

Polyethylene   A strong, lightweight thermoplastic that is used in food packaging, garment bags, garbage bags, construction coverings, and many other items.

Polyurethane   A widely used foam plastic that is extremely lightweight.

Polyvinyl Chloride (PVC)   A thermoplastic that is used in fishing and mosquito nets, shoe soles, floor tiles, and many other items.

Pop Rivets   Mechanical fasteners that expand when squeezed with riveting pliers; the excess metal drops off the head, leaving a formed head on the surface.

Portable Electric Drill   A power drill that can bore holes in different materials using variable speeds. This drill can install and remove screws, sand, file, polish, and strip finishes using different attachments.

Potential Energy   Stored energy that can be converted into other forms of energy.

Potentiometer   A variable resistor.

Pottery   Clay that has been heated at extremely high temperatures in a kiln to change it from a malleable to a rigid state.

Pouring   Using a material in a liquid state to make a final solid shape.

Primary Colours   Colours that contain no traces of other colours: magenta, yellow, and cyan.

Production   Fabrication of a design after it has been thoroughly tested and evaluated.

Properties   The qualities of a material, such as hardness, mass, water resistance, and so on.

Propulsion   The process of propelling or moving an object.

Prostheses   Also called artificial limbs, these are machines that are used by people who require hand, arm, or leg replacements.

Protective Gear   Protective clothing and devices worn when working with tools, machines, and materials (e.g., face shields, aprons). Each article of protective gear is designed to protect a different area of the body.

Prototype   An actual-size model of an object that is tested and evaluated before a design goes into production.

Pulley   A wheel that transmits force by means of a belt, rope, or chain passing over its rim.

Rabbet Joint   A mechanical joint that is an L-shaped groove cut across the end or edge of one part so that a second part will fit into it.

Radiant Energy   Light energy that is sent out from a source.

Radiation   The giving out of heat or light rays.

Radioactive   Material that is radioactive releases particles that interfere with the natural growth and health of other living things.

Rafters   The framing pieces that make up a roof.

Rasp   A coarse file that can cut curves and straight lines as well as sculpt wood. It is used for rough work, before finishing with a file.

Rearrange   To change the order.

Renewable   Resources that are renewable can be reused or replaced.

Repeat   To make multiples.

Researching In the design process, the stage of gathering and organizing information, drawing several designs, testing materials, and making a model of the design selected.

Resistor An electronic component that reduces the current of electricity in a circuit.

Resonance A reinforcing and prolonging of sound by the vibration of objects.

Reverse To change an idea to the opposite order.

Rip Saw A hand saw that is used to cut with the grain of wood.

Robots Complex machines that imitate human motion.

Roll The action of turning the length of an airplane to the left or right.

Rotary Motion A motion that goes around.

Routers Power tools that hold various cutting bits to shape wood.

Rubber Mallet A hammer with a head made of hard rubber that is used to bend sheet metal without denting it.

Sabre Saw A power saw that has variable speeds and can cut woods, plastics, and metals with a wide selection of replaceable blades.

Sanding Finishing with sandpaper to make smooth, clean surfaces and edges on a material.

Sapwood The outer part of a tree, under the bark.

Saw Pitch This describes the number of teeth/25 mm of a saw blade.

Saws Tools that have a toothed blade or disk that are used to cut different materials.

Screw A simple machine that is an inclined plane wrapped around a cylinder. Screws are also common mechanical fasteners that are threaded to fit into predrilled holes.

Scriber A sharp metal tool that looks like a pencil, which is used to scratch lines into materials.

Scroll Saw A stationary power saw that cuts woods, plastics, and metals using different speeds and types of blade. This saw can cut intricate shapes and can also accommodate a filing or sanding accessory.

Secondary Colours The colours (red, blue, and green) that are created by mixing two primary colours.

Second-Class Lever A lever in which the load is placed between the effort and the fulcrum.

Semiconductors Materials that allow precise control of electric current.

Separate To break apart.

Series Circuit An electric circuit in which two or more loads or controls are connected end to end. The same current flows through each of them, so there is only one path for the current.

Shafts The connecting axles between wheels in a belt drive.

Shape An area that is defined by a line.

Sheathing The outer covering of plywood that is nailed to the rafters of a house.

Siding The wooden, plastic, or aluminum exterior covering of a house.

Sills Wooden beams that are secured to the top of a house foundation to support the floor joists.

Sketching A rough form of drawing.

Smelting The process of melting and refining ores to make molten metal.

Snips Heavy-duty scissors that are used for cutting thick and hard materials, such as sheet metals and some plastics.

Softwood Wood that usually comes from coniferous trees.

Solder The metal alloy used as the "glue" in soldering.

Soldering A method of joining pieces of metal by using a melted metal alloy as the bonding substance.

Sole Plate Beams that are nailed on top of the floor around the perimeter of a house, to which the wall studs are attached.

Solution In a car battery, the acid substance that creates the electrical energy that activates the starter.

Solvents Chemical substances that are used for gluing, finishing, or cleaning a project or for thinning a finish. Varsol, plastic glue, methyl hydrate, and denatured alcohol are common solvents used in design and technology classrooms.

Sound A sensation of particles moving in waves.

Sound Energy Energy that is produced by vibrating objects.

Space   The area that surrounds an object or is contained within an object. Space is defined by the lines that encompass the object.

Spar   In a kite, the crosspiece that is attached at a right angle to the spine.

Spine   In a kite, the vertical piece of the frame that is attached at a right angle to the spar.

Springs   Devices that store potential energy when they are wound up or stretched.

Standardize   To make consistent or the same. In testing, to standardize means to use the same procedures in all tests.

Static Load   Anything that is part of a structure.

Steam Engine   An engine that uses burning coal to convert water into high-pressure steam. Steam engines are now obsolete and rarely used, but steam is used today to power turbines to generate electricity.

Stiffeners   Pieces of metal that are welded or riveted to each side of a girder to strengthen it.

Strip Heater   An electric heater that is used for heating sheets of plastic so they can be bent.

Structure   Anything that is composed of parts arranged together.

Studs   Upright pieces of wood that form the framework of house walls.

Subfloor   Boards or sheets of plywood that are nailed to the floor joists of a house, which provide the base for the flooring.

Substitute   To replace an idea with another.

Suspension Bridges   Bridges that use high-tensile steel wire to support the roadway.

Symmetrical Balance   A use of space where two sides of a shape or form are the same or are mirror images of each other.

Table Saw   Also called a variety saw or circular saw, this is a stationary power saw that has a circular blade that can cut woods and plastics.

Tail   The string attached to a kite that provides drag to stabilize the kite.

Template   A pattern that is used as a guide to trace the shape of an object onto a material, especially for making multiples.

Tension   A pulling force caused by stretching a material.

Texture   The feel and appearance of an object's surface.

Thermal Energy   Energy that is produced when the molecules of a material are heated and then begin to move.

Thermoforming   A process that involves heating a sheet of plastic until it softens so it can be forced into a mould.

Thermoplastics   Plastics that liquefy when they are heated and solidify when they are cooled.

Thermosets   Plastics that solidify under heat and cannot be made liquid again.

Third-Class Lever   A lever in which the effort is placed between the load and the fulcrum.

Thrust   The force that is created by the spinning propeller or escaping gases that draw or push an aircraft forward.

Torque   A twisting force that produces rotation.

Towers   Tall structures that are used for many purposes, such as supporting telecommunications equipment or carrying electrical wires.

Transistor   An electronic device that acts as a fast switch, switching current on or off or amplifying a small current.

Transmit   To pass along energy.

Trestle Bridge   A bridge that is built of many trusses or braced frameworks of wood or metal.

Triangle   A shape of structure that will resist any change of shape when forces are applied to it.

Trim   The decorative wood around windows, doorways, and the bottom of walls.

Truss   A triangular structure braced with supports that is used in construction.

Try Square   A tool that is used to test the squareness of a material or to draw lines that are perpendicular (90°) to an edge.

Universal Joint   A device for joining that transmits power from a shaft to another shaft that is running in a different direction.

Uranium   A radioactive material found in most rocks, soils, and rivers.

Veneer   A very thin layer of wood that makes up the top and bottom layers of plywood. Veneer is usually of a superior quality of wood to the other layers.

Vernier Callipers   A tool that combines callipers and a ruler to measure an object with great precision.

Vertical   A line that is perpendicular to a horizontal surface.

Vibration   A rapid movement of particles back and forth.

Warren Truss   A truss that consists of two single post trusses turned so that the bracing appears as a W.

Wedge   An inclined plane that moves to transmit force.

Weight   The downward force of gravity that acts upon the mass of an aircraft.

Welding   Bonding materials by using pressure or heat.

Wheel and Axle   A simple machine in which effort applied to the wheel causes the axle to turn with greater force than the wheel.

Wheels   Round frames that turn on a shaft in their centre, which are designed to reduce friction; for example, the front wheel of a bicycle.

Wire Cutters   Small snips that are used to cut wire, tin, and other soft metals.

Wood   A material that comes from trees and has a grain structure.

Wood Lathe   A lathe that is used to shape wood.

Workplace Hazardous Materials Information System (WHMIS)   A Canada-wide system to ensure that employers and employees are provided with information about the hazardous materials they work with on the job, to protect their health and safety. WHMIS compels employers to (1) place warning labels on containers of hazardous materials and (2) make available safety-data sheets with detailed information on hazardous materials (e.g., ingredients, fire and explosion data, health-hazard data, procedures for handling spills or leaks, and special precautions).

Worm Gear   A type of screw that is commonly used to transfer power from a motor to a wheel.

Yaw   The action of an airplane revolving around a central point.

Index